Diagnostics and Progno
of Aerospace Engines

Diagnostics and Prognostics of Aerospace Engines

BY RAVI RAJAMANI

INTERNATIONAL®

Warrendale, Pennsylvania, USA

400 Commonwealth Drive
Warrendale, PA 15096-0001 USA
E-mail: CustomerService@sae.org
Phone: 877-606-7323 (inside USA and Canada)
724-776-4970 (outside USA)

Library of Congress Catalog Number 2018948880
SAE Order Number PT-195
http://dx.doi.org/10.4271/PT-195

ISBN-Print 978-0-7680-9307-0
ISBN-PDF 978-0-7680-9539-5
ISBN-ePUB 978-0-7680-9309-4
ISBN-PRC 978-0-7680-9310-0
ISBN-HTML 978-0-7680-9308-7

To purchase bulk quantities, please contact: SAE Customer Service

E-mail: CustomerService@sae.org
Phone: 877-606-7323 (inside USA and Canada)
724-776-4970 (outside USA)

Visit the SAE International Bookstore at books.sae.org

contents

CHAPTER 1

Introduction

The propulsion system is arguably the most critical part of the aircraft. It is the single most expensive component. Because of its importance on most large programs, the engine is generally sourced separately from the airframe. Ensuring that engines operate reliably without major maintenance issues is an important goal for all operators, military or commercial. Engine health management (EHM) is an important piece of this puzzle and has been a part of engine maintenance for more than five decades. In fact, systematic condition monitoring was introduced for engines before it was applied to other systems on the aircraft. In this collection of articles from the archives of SAE International, we introduce the reader to a brief history of EHM, present some examples of EHM functions, both hardware and software, and outline important future trends.

The goal of EHM is ultimately to reduce the cost of operations and maintenance by catching problems before they become major issues, by helping reduce repair times through diagnostics, and by facilitating logistic optimization through prognostic estimates. A review of papers in this collection will show that the essence of these goals has not changed over time, even though the technology and design processes have grown in sophistication. It will also be evident that the full potential of EHM has yet to be realized.

Historical Perspective

Engine maintenance has been one of the more expensive components of airline operations and making it more affordable has been a goal of operators from the very beginning of commercial flight. The motivation for monitoring engines was to prevent any malfunction in the air. The paper from 1937 by E.O. Cooper [1] is included here for historical interest. He describes how radial engines were maintained by Pacific Airmotive Corp., an engine maintenance company that used to operate in Burbank, CA. Cooper describes the close relationship between original equipment manufacturers (OEM) and the operators, and how feedback from the latter drove many improvements. He talks about the relationships between the industry and federal regulators and the precursors to today's maintenance review boards. It is easy to see many of today's standard industry practices existing even then. Periodic inspections and scheduled engine overhauls were carried out to ensure preventive maintenance and increase flight safety. As a way of containing the costs associated with engine maintenance, practices such as on-condition inspection and maintenance arose. Condition monitoring was one way to further reduce costs associated with inspections since the ability to monitor equipment continuously reduced the need to inspect it periodically. It is interesting to note that many of the modern EHM practices were also developed at the request of airline customers.

All aircraft engines need sensors to control them, and many of these are linked to gauges in the cockpit. The earliest form of condition monitoring was manual, where pilots would monitor the gauges to ensure that critical parameters such as temperatures, did not exceed safe levels. Visual and audible alarms helped with this monitoring. Some gauges, such as those for oil pressure and temperature, were dedicated to safety functions. Later, accelerometers were added to monitor engine vibrations and detect rotor unbalance conditions. In contrast to the control sensors, these sensors were dedicated to monitoring the turbine's health.

With the advent of electronic controls in the 1980s, built-in-test (BIT) routines were added to the suite of tools used to establish the health status of various systems. BITs could be run during startup or they could be run continuously during operation. They monitored the condition of components such as sensors, actuators, etc., and were integrated into the engine control functionality. They provided a component's health status at a given point in time. Only with off-board data storage, or some means of recording and preserving data records, could we start to look at historical data and make some assessment about the progression of faults. This was done via data extrapolation which resulted in simple predictions. However, data trending is not straightforward in general, because raw engine data collected over different flights is not easily comparable due to varying ambient conditions, engines loads, etc. Before any comparison can be made, this gas-path data needs to be normalized and corrected to make all the data comparable. A pioneer in developing this technique of gas-path analysis (GPA) was L. Urban from Hamilton Standard [2].

GPA techniques were not just limited to commercial engines. In fact, the U.S. Air Force took a keen interest in GPA and applied it to their turbofan engines as well. As described by D.B. Edmunds [3], a similar performance-condition monitoring program was applied to the PW F100 engine in the mid-1970s. This was one of the first engines to be designed in a modular fashion so that the five different engine modules could be used and maintained independently. The goal was to reduce the Air Force's maintenance burden. It is interesting to note that a similar case for modularity was made for commercial engines in the 1975 paper by Bart and Nadler from Swissair [4]. They discuss how a set of disciplined maintenance practices helped the airline get the maximum usage from their engines, even when using partners such as KLM for maintenance. One key part of this discipline was the use of engine monitoring. They also distinguish between short-term and long-term monitoring; the former being the monitoring of components such as bearings and the oil system for immediate problems, and the latter being the long-term trending for finding developing issues.

In the collection of brief papers by Weitz et al. [5], which were presented at an SAE panel, we see the beginnings of EHM as early as the 1960s. Within this collection, the paper by J.J. Eden (from Trans-Canada Air Lines) shows how tracking the deviations of engine parameters from standards might help catch problems before they become major (Figure 1). This paper correctly points out that the sources of trend data are not just

FIGURE 1 Engine trend analysis [3].

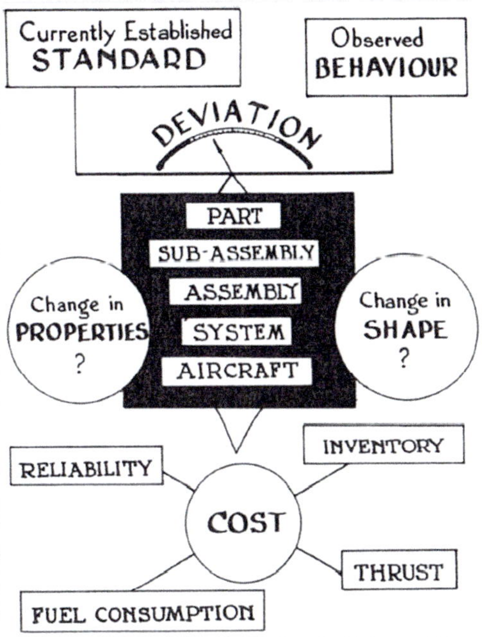

operational engines but also test rigs, test cells, and aircraft operating on the ground. Presumably, many of the records at this time were "stored" on paper or on magnetic medium, and analyses were done by hand after the data was transmitted to a central facility where a computer could be used for the analysis. Computers were still quite expensive and their use not as widespread as it is today.

In this collection [5], the paper by W.J. Cake [5] from Pratt & Whitney (P&W) discusses what the company provided airlines by way of support for trending turbine engine parameters and reducing in-flight issues such as shutdowns. Cake also mentions how the military issued an Operational Support Requirement (OSR-322) in 1958, where they put out a request for engine monitoring capability in the form of a Jet Engine Analyzer System. The overall goal of the OSR was to "provide a jet engine analyzer system, consisting of airborne and ground components, to measure jet engine performance, diagnose and isolate troubles, and aid in predicting engine malfunctions and failures." A request way ahead of the times, it would not be until about a decade later that anything close to this would be developed.

The paper by Trans World Airlines' C.A. Fisher gives an insightful history of trend monitoring for jet engines on the Boeing B707, including details of how magnetic tapes were used to continuously record data from the P&W engines. Reviewing the results of engine trending, Fisher concluded that while it definitely affected delays, in-flight shutdowns (IFSDs), and unscheduled engine removals (UERs), the main benefit was in identifying inefficiency trends (i.e., increased fuel burn) and in the ability of the airline to correct such problems early. This helped save fuel costs which were a large part of the operating expenses for the airline. In fact, according to the author, "TWA is convinced that trend analysis is an essential part of the maintenance program for all large jet powered transport aircraft." Of particular note in this collection is the final paper by E.J. Briggs—from the Federal Aviation Administration (FAA)—who discusses the future of engine trending. He has some amazing insights into how, in the future, trending might be used to eliminate scheduled overhauls and have them be triggered by the observed condition.

FIGURE 2 Performance calculator nomogram from P&W (courtesy of Michael Kennedy).

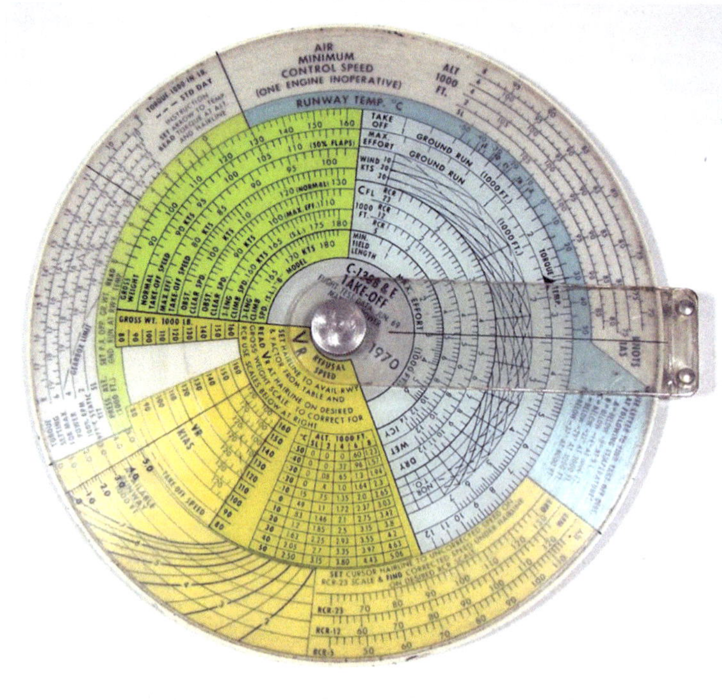

Michael (Mike) Kennedy, who used to manage P&W's performance monitoring group in the 1990s, and personally witnessed all the changes in EHM technology from manual hand-plotted charts to the current era of cloud computing, indicates that the original motivation for performance monitoring really came from the airlines, especially the European ones. They realized the benefits that could be derived from trending parameters and got the engine OEMs to help them set up EHM procedures and aids. An example of the latter is an old circular slide rule (also called a nomogram) to calculate corrected parameters on the fly (Figure 2). In the era before computers, airlines often had flight engineers on board, who could help with such calculations. But no direct calculations were generally needed for engine monitoring. Cockpit indicators readings were recorded by hand once the flight had settled into its cruise phase. This essentially meant that the aircraft had achieved steady flight, i.e., no throttle movement for a given amount of time—say about 3 min. These parameters were engine pressure ratio, speed, fuel flow, exhaust gas temperature (EGT), and possibly oil pressure and temperature, and vibration levels on some aircraft. These would be handed over to the maintenance crew after the flight. All these pieces of paper would be analyzed on a periodic basis to track engine performance. This process was clearly time consuming and prone to errors, so airlines were motivated to look for automatic recording equipment to be installed in the aircraft. In response to this need, companies such as Teledyne, Lockheed, Allied Signal, Hamilton Standard, and SFIM (a French company now part of Safran) developed data acquisition equipment with technology similar to crash recorders. The recording media consisted of magnetic tapes which were removed on a periodic basis and transcribed on the ground to facilitate performance tracking. At first, airframers were not the ones providing these monitoring systems as part of the original equipment contract; airlines did this as supplemental equipment for which they obtained the necessary certification. Later, in the 1980s, as the airframers realized the benefits of data recording for diagnostics and performance trending, they started offering the equipment, first as options and later as standard equipment. As the aviation world transitioned from analog to digital, the sophistication of data acquisition equipment grew with increasing capacity and speed. With the advent of broadband connectivity, this is now reaching the state where data throughput, or its lack, is not the limiting factor in EHM. If anything, there is too much data and what we need are smarter algorithms to make better sense of all of it.

Another trend in the evolution of EHM, according to Kennedy, was the miniaturization of the computer. As the large mainframes gave way to the desktop personal computer (PC), engine OEMs began to develop EHM programs that could be run on PCs. These were distributed to the airline customers as a means of supporting the aftermarket spare parts business. The erosion of the parts business by third-party Parts Manufacturing Approval (PMA) programs in the beginning of this century made the engine suppliers look at EHM as a business opportunity independent of spare parts. Consequently, EHM is now a service that operators must pay for, generally on a fixed-price basis (i.e., per engine flight hour), often as part of a larger fleet management program. Most engines sold today are covered by some form of a long-term service agreement.

Historically, aircraft maintenance practices were very much based on scheduled inspections and overhauls. In the 1960s, the operators realized that this was expensive and did not necessarily enhance the reliability of the airplane. They formed the Maintenance Steering Group (MSG) and, with the Boeing B747 as the target program, ushered in new guidelines for maintenance that was based on a rigorous scientific understanding of different failure mechanisms [6]. Reliability-centered maintenance (RCM) practices were incorporated into MSG, as were more on-condition maintenance practices. Moving away from hard-time maintenance and toward more condition-based practices has helped reduce costs associated with maintenance in the aviation world [7]. For an interesting look at the history of condition-based maintenance (CBM), the reader is directed to a companion collection of SAE papers [8] in the same series. Engines have been a major beneficiary of this trend. Smith, DeHoff, and Huffman describe a program that the Air Force conducted with General Electric (GE) to monitor GE engines with on-board trending [9]. This was done to support the Air Force's RCM philosophy for engine maintenance. In this paper, the authors also describe some success with trending non-steady-state data taken during each takeoff. According to the authors, this flight regime is quite repeatable for military engines and is amenable to meaningful trending after standard corrections.

EHM, as it has evolved to the present time, consists of a few salient features. Subsystem or component EHM is the monitoring of distinct subsystems using dedicated sensors and accompanying analysis. System-level EHM relies on an entire set of distributed sensors to monitor specific issues, or the analysis of the entire engine based on more extensive models and algorithms. Regardless of the type of EHM, some common elements of all these functions are data acquisition, secure data transmission, storage and retrieval, analysis, and display or annunciation. This entails both on-board and off-board elements, which is what makes EHM so unique, requiring a more holistic approach during the design, testing, and operational phases.

Over the years, the goal of EHM has always been to reduce the cost of operations and maintenance by catching problems before they become serious issues. For example, operators want to avoid IFSDs because they impact safety as well as other aspects of flight, such as extended operations (ETOPS). Predicting failures can help lower these rates. Similarly, UERs have a negative impact on all metrics, such as availability, delays and cancelations (D&C), and maintenance cost. Turning UERs into scheduled maintenance visits is much better, and EHM systems can help do this. If advanced diagnostics can identify and isolate problems beforehand, less time is spent in the repair shop and hence reduces the maintenance turnaround time, thus helping return the engine to service sooner.

Another important aspect of EHM that is not emphasized too much is its effect on optimizing the aftermarket supply chain. Prognostic signals can allow prepositioning of parts, the correct routing of aircraft to the appropriate airports with maintenance facilities, and substitution of assets when appropriate to avoid other costly procedures. All this will help increase the overall effectiveness of the maintenance work, helping to lower an expense that operators have much more control over, when compared to other costs such as fuel or labor [10].

System Architecture

All EHM systems are configured with on-board and on-ground elements. How much of the processing needs to be done on board vs. on the ground depends on many factors, such as available processing power, availability and bandwidths of transmission infrastructure, business processes including contracts with customers, criticality of data needs for maintenance, ability to easily update software, etc. A good reference for EHM in general is the Aerospace Recommended Practice (ARP5120) that has been authored by the SAE Technical Committee E-32: Aircraft Gas Turbine Engine Health Management System Development and Integration Guide [11]. In fact, the E-32 committee [12], established in 1975, is one of the earliest groups to write and publish documents on different aspects of propulsion health management and is a good resource for information and standards about his subject. A related committee is HM-1, which publishes documents on vehicle health management [13].

Data security plays a role in the selection of the architecture as well. For example, in a military application, where the consequences of a data leak are very high, more importance will be given to reducing unnecessary data transmission. This might mean that more analysis is done on board, as compared to a civil application where this burden may be shifted toward ground-based systems. Another reason to choose the latter would be because computational resources are much more expensive if they are on board as opposed to being on the ground. Such trades are generally addressed before the architecture is finalized (Figure 3).

FIGURE 3 The different phases of the EHM system.

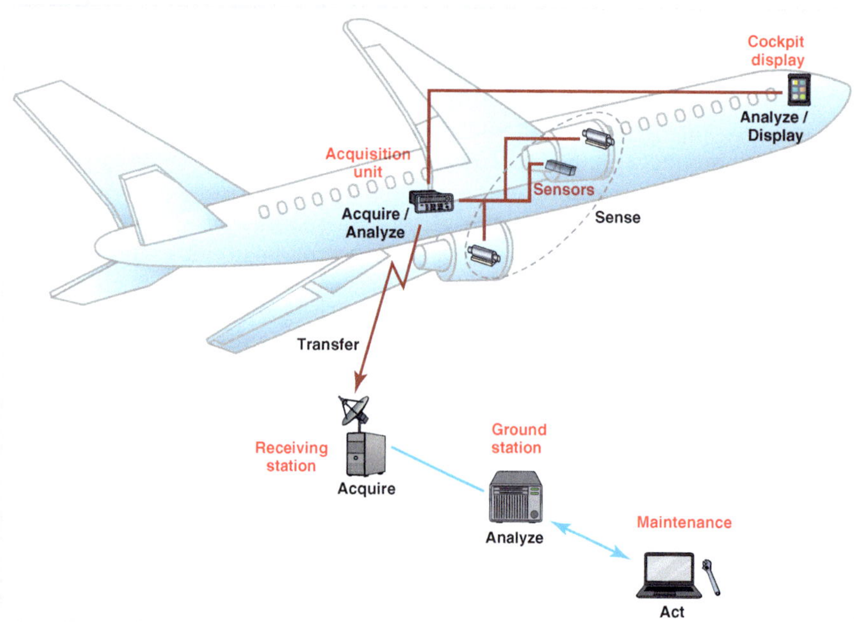

These considerations lead to architectural decisions about the proper way to transfer data: within the engine, between the engine and the airframe, and from the airframe to the ground. The most popular way to do the transfer to the ground is through the Aircraft Communications, Addressing and Reporting System (ACARS). According to Wikipedia [14], the motivation for the development of ACARS was to accurately record and report when certain flight phases where achieved, something that was done manually in the past. As the article points out, "ARINC introduced the ACARS system in July 1978, as essentially an automated time clock system." Piedmont Airlines was one of the pioneers in installing ACARS communication equipment on all its planes, according to M. Kennedy (and independently confirmed in the paper by Kaiser and Vernaleken [15]).ACARS proved to be a boon for the condition monitoring community, because key reports could now be generated in flight, downloaded automatically, and processed by off-board software even while the aircraft was still in the air. ACARS is a very expensive way to transfer data however, and while many alternatives are now available, they are still not in extensive use. Cell modem and Wi-Fi at the gate are two relatively inexpensive transmission modalities that sacrifice immediacy for lower costs. Satellite transmissions are becoming cheaper as well, and as in-flight entertainment systems force airlines to expand their broadband connectivity, health management systems will benefit. Because of lower storage costs, engine OEMs are increasingly designing EHM systems that can collect and store high bandwidth data. This has led to the need for superior transmission infrastructure so that all this data can be exploited for better diagnostics and prognostics.

The design of a good EHM system requires a careful systematic approach, beginning with requirements and concluding with validation in field trials. As with any other aerospace system this takes time, knowledge, and experience. A good reference for learning about the entire process is the paper by Snecma's W. Bense [16]. Interestingly, Bense describes the process of design following the guidelines outlined in the Open System Architecture for Condition-Based Maintenance (OSA-CBM) standard [17]. The paper is very comprehensive and describes all aspects of diagnostics and prognostics as it is practiced within Snecma.

Hardware

The most popular form of jet engine used on commercial aircraft is the high-bypass gas turbine, called a turbofan. Figure 4 shows a two-spool machine, which is its most popular variation [18]. Of the three major jet engine OEMs, the two based in the United States—GE and P&W—make engines with this architecture, while Rolls-Royce, headquartered in the United Kingdom, makes engines mainly with a three-spool design. By adding an intermediate stage, the three-spool design derives some efficiency advantages at the cost of additional mechanical complexity. More recently, P&W has developed a geared turbofan engine that allows the fan to rotate at a slower speed than the low-pressure shaft, thereby increasing overall efficiency, in a design that is a compromise between the two- and the three-spool architectures. Figure 4 also shows axially designed modules, which is generally the case with larger engines. Smaller engines might have centrifugal compressors and turbines to save space, but the principle of operation is still the same. In all high-bypass designs, most of the thrust is derived from the fan pushing air through the fan duct. A portion (roughly 7.5% to 15%) of the total airflow is directed to the core of the engine where it passes through the (low- and high-pressure) compressors and the combustor. Here, its temperature and pressure are increased by burning fuel. This increase in energy is used to power the turbines which, in turn, power the compressors and the fan, thus closing the cycle. Most large civil aircraft are powered by turbofans.

FIGURE 4 Two-spool high-bypass aircraft engine with major components and station numbers [18].

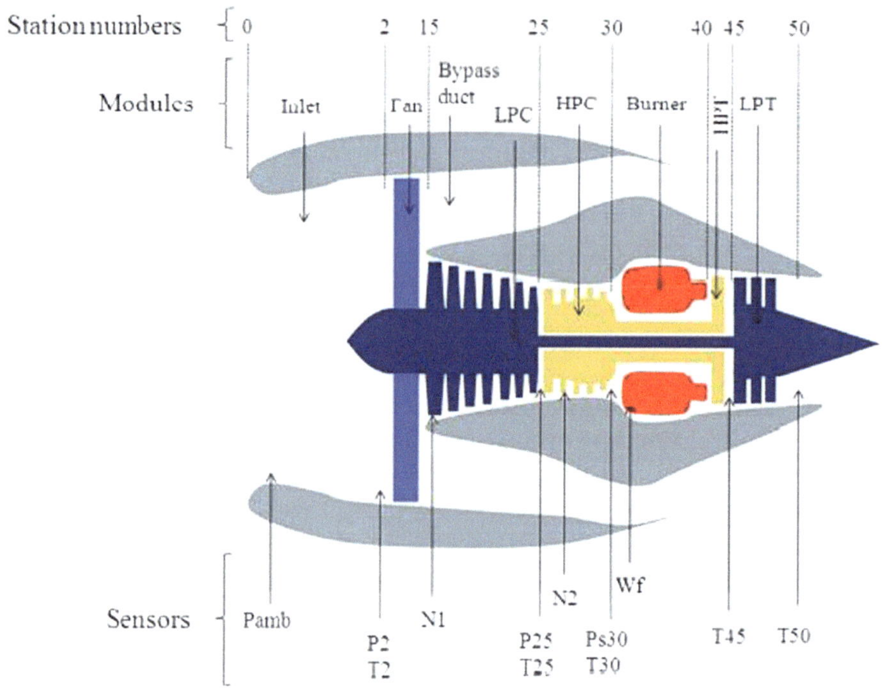

Older military jets might use turbojets, where all the air goes through the core, whereas modern military jets use low-bypass turbofans. Smaller aircraft are often powered by turboprops which are essentially turbojets that drive free turbines attached to propellers. This is the same arrangement found in helicopters, where the main rotor and the tail rotor are driven by transmissions powered by one or more gas turbines. Smaller, general aviation (GA) category aircraft are often powered by internal combustion reciprocating engines. This book does not discuss reciprocating engines.

Regardless of the type of gas turbine, there are a few sensors that are always present. These are needed for the control of the engine but can also be used for health management functions. In fact, the FAA mandates the use of certain sensors in a power plant to ensure safety. A comprehensive list is available on their electronic Code of Federal Regulations (CFR) site [19], in Part 25.1305 titled "Powerplant Instruments."

In addition to sensors required for controls and for ensuring safety, modern engines have additional sensors that are installed specifically to provide EHM functions or to affect finer controls. In the following discussion, we will be referring to Figure 4; the necessary extension to a three-spool machine should be straightforward.

The very basic sensors needed to operate a gas turbine are inlet temperature and pressure (T2 and P2), spool speeds (N1 and N2), exhaust temperature and pressure (T5/T50 and P5/P50), and fuel flow (Wf). A few safety-related sensors for oil, fuel, and fire are installed as well. Over the years, these have come to include: oil quantity, pressure and temperature, fuel quantity and pressure, engine fire indicators, vibration monitors, ice-protection and thrust-reverser status indicators. More sophisticated controls began to use combustor inlet conditions (T3/T30 and Ps3/Ps30), especially to avoid compressor stall situations. A few military engines use pyrometers to measure turbine inlet temperature (T40), and at least one commercial engine briefly introduced this measurement, but it seems in this case the sensor was not reliable enough to be useful for engine controls or even for EHM purposes. Ground-based turbines have been more successful in this area. The use of pyrometers to measure hot gas-path temperatures is not a new idea [20]. Modern engines have sensors that measure intermodular conditions as well, such as T25, P25, and T45. Additionally, some form of automated oil-debris monitoring (ODM) system is becoming common on modern engines. The latest military engines have additional dedicated EHM sensors, such as blade health sensors [21, 22], inlet and exhaust airflow debris monitors, and oil-condition monitors [22], but these are still so rare that they will not be covered in detail here.

Monitoring the combustion process is something that has always attracted the gas turbine community, because it is the heart of the entire Brayton thermodynamic cycle that the gas turbine operates on. The physics of the fuel atomizing, mixing with air, igniting, and sustaining a flame, in a resonating chamber, are exceedingly complex. Therefore, much of today's technology has been built up over the years through extensive experimentation. The goal of combustion is to maintain a flame that will exhaust all available fuel and achieve the highest sustainable temperatures without producing any more polluting gases than absolutely necessary. While doing all this, the flame should not excite any resonant modes that might destroy the combustor through high-pressure fluctuations. Adding pressure and temperature sensors to monitor the combustion process would help, but because of the extremely hot conditions, it has been very difficult to install sensors and use them for monitoring internal conditions during normal operations. However, test instrumentation is routinely installed during testing, so that the control strategies can be tuned before the system is deployed.

Mechanical diagnostics was one of the first applications for EHM in an engine. The oil circulation system is extremely critical to engine operation, providing not only lubrication but also heat rejection in bearings and gearboxes. Any major leak in, or malfunction of, the oil system would be catastrophic. Oil pressure and oil temperature sensors allow the pilot to constantly monitor the vitals of the lubrication system and detect problems before they become unmanageable. In addition to these sensors, magnetic plugs functioning as chip detectors are used on the oil scavenge lines to identify bearings that might be malfunctioning. Most bearing material is ferromagnetic, and chips carried by the oil have a high probability of being captured by the plugs. Regular inspection of these "mag-plugs" might reveal problems with bearings, such as spalling and pitting, that could prevent further damage and a catastrophic engine failure. Additionally, filters are generally equipped with bypass conduits that allow oil to flow around them if they are clogged. This could also trigger a warning in the cockpit that would force an inspection and corrective action.

Having to inspect plugs on a regular basis is a chore that costs time and money. Many new devices have been developed and deployed to reduce this maintenance burden. Chief among them is the electronic chip detector which provides a method of triggering an alarm when a sufficient number of chips have collected on the tip—often by the means of bridging a gap between conductors, thereby closing a circuit. An advanced version of this system, made by Tedeco (now a division of Eaton Aerospace), is called a quantitative debris monitor (QDM) which uses electronics to count the number of chips that are attracted to the magnetic chip detector. The paper by Higgins and Crow [23] describes the QDM installed on the General Electric GE90 engine, which they claim was the first FAA-certified oil-condition monitoring system. The QDM is designed to count individual debris particles and their distribution relative to a target particle weight threshold. The system must be calibrated on the type of bearing material that is likely to be seen. The authors claim that in the two years of service with the GE90 the QDM system performed with nearly 100% accuracy and a minimal false alarm rate. The QDM employs a cyclonic separator for oil, air, and debris. The debris particle is captured by a magnet, but in so doing it passes near the coils of an electromagnet where it generates a pulse that is recorded by the sensor electronics. These pulses count the total number of particles flowing in the lubrication system and drive maintenance if a certain threshold is exceeded.

A competing technology used by many military engines and ground-based gas turbines is an inline ODM system made by Gastops that counts particles as they flow through it [24]. This is an inductive device as well but can distinguish between ferrous and nonferrous metallic particles. This is important because rotating components may not necessarily be made of ferrous material. According to their website, this ODM system is installed on a number of aircraft engines and ground-based equipment [25].

The ODM is a key condition monitoring tool, but there are enough sensors in the lubrication system to be able to model and analyze it at a system level, which is the approach taken by M. Ma from the China Flight Testing Establishment [26]. The author proposes building a data-based model of the entire oil system (represented by oil pressures and temperatures in the system) using actual full-flight data (temperatures and pressures, as well as flight conditions such as throttle position). The model, which is an artificial neural network, predicts these oil parameters, which the author then compares to the actual parameters and uses the deviations to assess the condition of the system. It is an intriguing idea, but it is not entirely clear how the author distinguishes between steady-state and transient conditions because this will have a profound impact on the accuracy of the neural network. Nevertheless, it is an interesting approach and it is practical as demonstrated by its use on real data.

While oil-system monitoring is an excellent example of a novel engine health monitoring technique for mechanical systems, perhaps the most important EHM function is provided by monitoring the vibration sensors on the engine. According to Greaves and White [27], vibration monitoring on engines was introduced in the 1960s and became more widespread after the FAA mandated monitoring rotor "unbalance" via Federal Aviation Regulations (FAR) 25.1305 (d) (3) in 1974. Earlier systems were analog in nature but all the modern airborne vibration monitoring (AVM) systems are digital with piezoelectric accelerometers, data acquisition systems with sophisticated algorithms to do much of the preprocessing of the raw vibration signals.

Typical accelerometers for commercial applications have bandwidths of less than 5 kHz, while newer military applications may use sensors that have bandwidths an order of magnitude higher. A lot more can be discerned at these higher frequencies, with the right algorithms. Watson et al. [28] look at high-frequency signals to pick out faults in gear teeth. These are experimental setups but show promise for real-world applications.

Software and Analytics

Analytical methods have been just as important for engine diagnostics as the use of sensing systems. In fact, for current engine programs, these techniques have seen a lot more advances. As early as the 1970s, engine OEMs began developing GPA algorithms to analyze engine performance and discern engine issues. Pioneering work in this area was done by L. Urban at Hamilton Standard [2]. This work established the GPA program for P&W, which today is codified in a program called Advanced Diagnostics and Engine Management (ADEM), also offered as part of their EngineWise maintenance, repair, and overhaul (MRO) program [29].

Soon after P&W developed its condition monitoring program, GE developed a similar program [30] which has now evolved into PHM+. Rolls-Royce has its flavor of an engine-monitoring program as part of the TotalCare package which runs, partly on the cloud, with the help of Microsoft Azure [31]. Smaller companies had developed their own engine

FIGURE 5 Information flow for engine diagnostics [34].

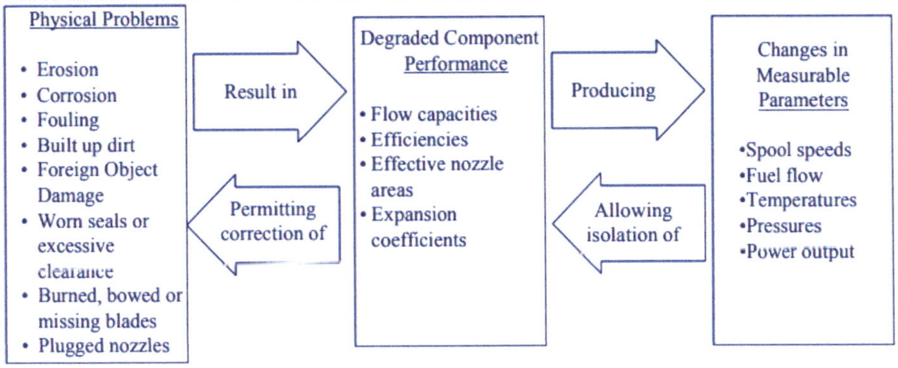

management programs, such as Pratt & Whitney Canada with WebECTM, and Honeywell with EHM. Both are now run by CAMP Systems Inc. [32]. The programs from the larger companies go far beyond engine monitoring. In fact, depending on the level of service bought by the customer, they include parts replacements and overhauls as well. A term that is used in the industry to describe this is "power by the hour," which was originally coined by Bristol Siddeley Engines Ltd., a British aeroengine manufacturer later purchased by Rolls-Royce [33]. The idea is that the engine OEMs are really selling "thrust," rather than just an engine. The risks associated with maintaining and ensuring reliability is passed on to the OEM rather than the operator, in return for which the OEM gets paid a fixed fee per flight hour. This ensures a predictable revenue stream for the OEM and an equally predictable expense outlay for the operator. In such a business arrangement, the use of EHM becomes easy to justify because it helps to lower costs and mitigate risk for the maintainer.

The basic philosophy behind GPA is laid out in Figure 5 from a NASA paper [34]. Any degradation of the engine, either an abnormality or normal age- or usage-related deterioration, will lead to changes in some internal "performance health" indicators—essentially flows and efficiencies—which will show up as changes in measured parameters. Once the differences between these measured parameters and the estimated "healthy" parameters become large enough to be statistically meaningful, the GPA system will raise a flag, which would then result in further analysis to isolate and correct the problem. Traditionally, GPA is a steady-state calculation, using data from cruise, when all engine conditions are relatively stable. Correcting all parameters to standard day condition allows the differences (also called "deltas") to be compared over time, thereby facilitating parameter trending over multiple flights. Even slow changes in system characteristics can be discerned over time. Or one engine can be compared with another or with the rest of the engines in the fleet, providing more insight into any specific issue.

A classic GPA performance monitoring parameter is the EGT margin at takeoff. As a new engine accumulates operating hours, it is inevitable that the compressor and turbine blades will foul up. This effectively reduces the efficiency of the compressor's pumping action and the effective flow area of the turbine. The net result is that for the same thrust, the EGT is higher for an older engine when compared to a new engine, and the difference between this and a maximum "allowable" limit is called the takeoff EGT margin. To ensure that this EGT margin can be compared over various different ambient conditions, the calculation involves a correction to standard conditions, very similar to what is done for other GPA calculations. When this number reaches zero, the engine needs to be serviced so that the lost EGT margin can be recouped. This is usually done with a water wash, either on the wing or in a shop, but might involve more extensive repairs based on other inspections.

Software associated with EHM has never had to be approved by the regulators, unlike safety critical software that is responsible for controlling the engine. This is because health management has never been linked to safety; engines are designed to be safe without the use of EHM. Additionally, much of the software associated with EHM runs on ground-based computers, and there never has been a good way of evaluating ground-based systems, unless they are dedicated for tasks such as navigation. In the latter case the Radio Technical Commission for Aeronautics (RTCA) has published guidelines for developing such software [35] based, to a large extent, on their recommendations for airborne software [36]. There is some effort within the SAE Technical Committee E-32 to develop guidelines for EHM software. This is needed now because OEMs are realizing that EHM software can have safety implications.

A case in point is engine ETOPS, where aircraft operating far from a safe landing location (such as over the ocean) need to be assured of making it safely back to land in case one engine becomes inoperative in flight [37]. In case of an IFSD, the other engine needs to be ramped up to a maximum continuous thrust (MCT) level that will allow it to power the aircraft for a given amount of time (which depends only on the design limits of the aircraft). The way an engine is assessed to be able to make MCT is to either test it on a periodic basis with ground runs or to analytically correlate it to a tracked parameter such as EGT margin, which is already an EHM function. Today, even if an EHM system is used to decide whether an engine can be dispatched for an ETOPS flight, the decision ultimately rests with a competent human being. If this ever becomes an automated process, it will clearly have safety implications. Then the design assurance level of EHM software will have to be considered to make sure that it is reliable enough to support such a function. This will be the next evolution of EHM software, as it becomes a more important part of establishing the airworthiness of the engine.

Engines have some of the most expensive life-limited parts (LLPs) in an aircraft. These, as their name suggests, are only allowed to operate for a given number of cycles or hours. All rotors (that hold the blades) are life limited. The OEMs use extensive modeling and analysis as well as limited testing to arrive at these life limits. A number of assumptions go into these calculations, such as aircraft loading, ambient conditions, runway location, etc., and the ultimate life limits are always conservative to ensure safety. It has been shown that a majority of the LLPs reach their stated lives much earlier than they should; hence they are replaced when they still have useful life left in them. This life-wastage can be avoided if the same OEM models can be used to estimate actual life lost each time an engine operates, instead of using the same conservative number for every flight. With such a usage-based lifing (UBL) philosophy, these expensive parts could operate longer on-wing before they are replaced. This methodology will be relatively easy to design and implement because it is based on existing mechanical design philosophy; no certification documentation needs to be changed. What needs to change, however, is the method the operators use to estimate remaining useful life (RUL) after every flight based on the actual life lost during that flight. Military engines are ahead of the game here, with modern jets such as the Gripen, Eurofighter, and Joint Strike Fighter (JSF) using UBL to get more operational life from their LLPs when compared to older aircraft. The Onboard Life (Consumption) Monitoring System (OLMOS) program for the Eurofighter [38], the Life Tracking Systems (LTS) for the Gripen [39], and the Engine Life Monitoring System (ELMS) for the Rafael [40] have been in operation longer than their U.S. counterparts, but this is a methodology that is clearly going to be adopted universally because of the enormous savings that can be realized. Currently commercial engines have a more deliberate process for extending operational life called derating. Here, the pilot can manually reduce power while taking off based on the takeoff weight, ambient conditions, etc., and this allows the engine's life debit to be "de"-rated for that flight.

An automatic system will clearly be more useful and more commonly used. But, before that can happen, authorities need to be convinced that the operators know what they are doing, not only with the appropriate algorithms but also with the software.

Future Trends

There are a number of general trends in both hardware and software to advance technology and implementation of EHM. New sensors and new applications for existing sensors are constantly being tried. But better models and model-based techniques using existing sensors give us a more immediate way of enhancing EHM capabilities. For example, most engine GPA is based on steady-state data. The algorithms essentially compare expected steady-state parameters against measured parameters to detect and identify faults. However, data gathered from transient operation of the engine should contain a lot more information about the health of the engine, and if some dynamic model-based scheme could be developed to exploit this information, then diagnostics could be carried out much faster and more reliably. The downside of course is that the number of false alarms might increase. Using dynamic data to detect problems is not a new idea, even in the context of gas turbines. An early application of this technique to a gas turbine was the use of a Kalman Filter along with a dynamic model to detect leaks and blockages in the cooling circuits of a GE H-machine [39]. The algorithm was able to detect and alert on critical faults much faster than any steady-state system could. This was used on an industrial gas turbine. Similar techniques are starting to be used for aircraft engines as well.

Consider the ideas put forth by Volponi and Tang [42] where they propose to use dynamic data to increase the capabilities of GPA algorithms. The authors propose capturing data from the entire flight and using it, along with a dynamic model, to accomplish dynamic GPA. They also propose a methodology of comparing information from multiple engines which is very innovative, since most diagnostic algorithms are self-contained and only analyze the engine they are monitoring. Incorporating meta-analysis can help with correlated faults or eliminating extraneous disturbances if they affect all engines similarly.

The other big trend is the integration of EHM and controls. This has happened quite readily with hardware, where many modern aircraft (especially on the military side) have combined electronic engine controls and EHM hardware. This is efficient because both controls and EHM share the same inputs, and having the same enclosure leads to some weight savings. But this has some disadvantages as well, especially with development costs, maintenance protocols, etc. However, the trend we are talking about here is more on the functional side. Take for example the paper by Behbahani et al., from the Air Force Research Laboratory (AFRL) [43]. The authors discuss the integration of various functions in a future military aircraft with a focus on the engine. They envision a future where PHM is used to assess the health of the aircraft, and the controls are used to appropriately modify the controls to accomplish the mission. But the paper goes beyond this. It looks at various subsystems such as the thermal management and the power subsystems and describes how distributed controls can be used to integrate them. The goal is to optimize the operation, so that any mission is accomplished with the expenditure of the least amount of resources, and the controls can adapt to the estimated health condition. Figure 6 shows how the authors envision the integrated system.

Clearly, aircraft maintenance in the future will be different from the way it is done today. Individual subsystems will be smarter and able to assess their own health on a continuous basis. These capabilities will be consolidated by the vehicle-level reasoning system to make intelligent decisions about maintenance. With the PHM system integrated with the controls

FIGURE 6 Integrated future vehicle [43].

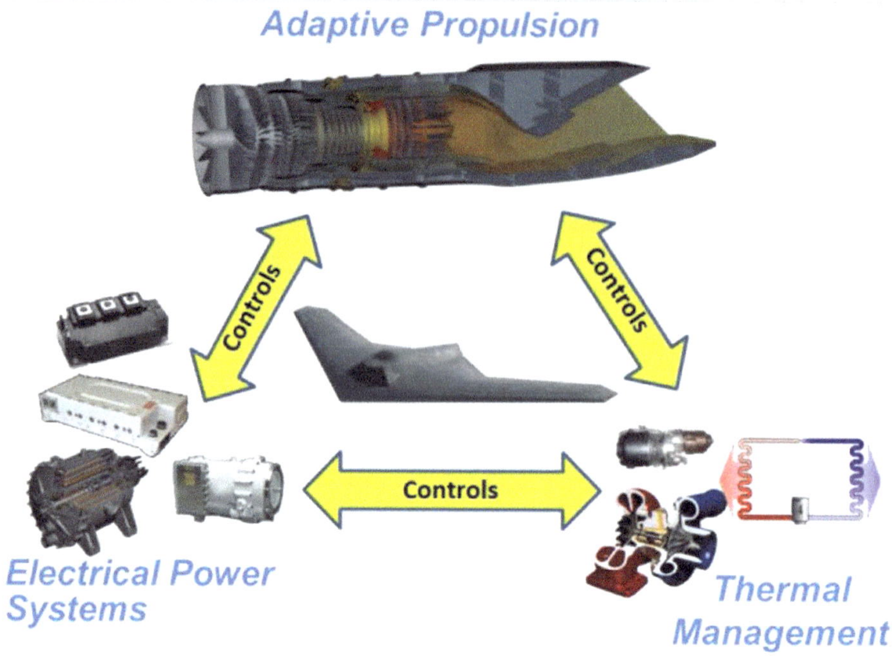

system, it is inevitable that control functions will be adapted depending on system health. Not only that, with a seamless integration of on-board and off-board systems, health indications from the vehicle will automatically trigger actions in the aftermarket logistics infrastructure. It is not inconceivable to envision a future where parts are automatically ordered, overhauls planned by optimization algorithms, and even scheduled operations rerouted to accommodate different vehicles' health conditions. Research has shown that with an integrated system using advanced PHM concepts, asset availability can be increased, and part usage decreased without any compromise in service or safety [44], but there is a difference between demonstrating this in theory and translating it into practice. This will require not only innovative designs but a whole new regimen of testing and certification so that safety is never compromised. This means that the level of collaboration and information sharing between all players in the industry around the world will have to grow. With electric propulsion and other innovative business models like urban air-taxi services, the future of aviation is exciting. And, not unlike the field of medicine, the technologies and innovations related to health management will continue to advance and provide a fertile area of research and development for generations of propulsion engineers.

Conclusions

In this collection of papers from the archives of SAE, we have charted a history of the field of EHM. SAE has been there from the very beginning, and we trust that many of the papers included here will provide an interesting insight into how some of the key technologies associated with EHM developed and advanced. EHM has progressed through the growth of hardware and software solutions. Many challenges remain, and it is expected that these will provide future engineers with many opportunities for exciting work.

Acknowledgments

I am grateful to Mike Kennedy for taking time out of his busy golf schedule to discuss the history of EHM with me in a manner only an experienced veteran in the industry could. An unusual cold snap in Florida this winter probably helped get Mike indoors long enough to field a few calls! And I would be remiss if I did not thank my editor, Monica Nogueira, for her constant help in the making of this book.

Abbreviations

ARP - Aerospace Recommended Practice
BIT - built-in test
CFR - Code of Federal Regulations
CBM - condition-based maintenance
EHM - engine health monitoring
ELMS - Engine Life Monitoring System
ETOPS - extended operations
FAA - Federal Aviation Administration
FAR - Federal Aviation Regulations
GA - general aviation
HM - health monitoring
IFSD - in-flight shutdown
IVHM - integrated vehicle health management
JSF - Joint Strike Fighter
LTS - Life Tracking Systems
MCT - maximum continuous thrust
MSG - Maintenance Steering Group
MRO - maintenance, repair, and overhaul
NASA - National Aeronautics and Space Administration
OEM - original equipment manufacturer
OLMOS - Onboard Life (Consumption) Monitoring System
PHM - prognostics and health management
RCM - reliability-centered maintenance
RTCA - Radio Technical Commission for Aeronautics
RUL - remaining useful life
TAN - Total Acid Number
TC - type certificate
TWA - Trans World Airlines
UER - unscheduled engine removal

References

1. Cooper, E.O., "Aircraft Engine Maintenance," Presented to Southern California Section of the Society of Automotive Engineers, https://doi.org/10.4271/370035, April 1937.
2. Urban, L.A., "Gas Path Analysis Applied to Turbine Engine Conditioning Monitoring," *Proceedings of the AIAA/SAE 8th Joint Propulsion Specialist Conference*, New Orleans, LA, November-December 1972.

3. Edmunds, D.B., "Modular Engine Maintenance Concept Considerations for Aircraft Turbine Engines," *Aerospace Meeting*, Los Angeles, CA, November 1977, doi:https://doi.org/10.4271/770991.
4. Bart, H.W. and Nadler, A.B., "Maximum Installed Engine Utilization through Disciplined Health Monitoring," *Proceedings of the Air Transportation Meeting*, Hartford, CT, May 1975, doi:https://doi.org/10.4271/750612.
5. Weitz, G.H., Stahr, R.S., Eden, J.J., Cake, W.J. et al., "Aircraft Gas Turbine Engine Trend Analysis," *Proceedings of the SAE World Congress & Exhibition*, January 1964, doi:https://doi.org/10.4271/640611.
6. Nakata, D., "An Introduction to MSG-3," *Proceedings of the Aerospace Congress & Exposition*, Long Beach, CA, October 1984, https://doi.org/10.4271/841481.
7. Nowlan, S.F. and Heap, H.F., *Reliability-Centered Maintenance*, (Los Altos, CA: Dolby Access Press, December 1978).
8. Rajamani, R., *Condition Based Maintenance in Aviation: The History, the Business and the Technology*, (Warrendale, PA: SAE International, 2018).
9. Smith, C., DeHoff, R., and Huffman, M., "Developments in Performance Monitoring and Diagnostics in Aircraft Turbine Engines," *Proceedings of the Aerospace Congress & Exposition*, Anaheim, CA, October, 1982.
10. SAE International, "Determination of Costs and Benefits from Implementing an Engine Health Management System," ARP4176, February 2013.
11. SAE International, "Aircraft Gas Turbine Engine Health Management System Development and Integration Guide," ARP5120, March 2016.
12. SAE International, "E-32 Aerospace Propulsion Systems Health Management," https://www.sae.org/servlets/works/committeeHome.do?comtID=TEAE32.
13. SAE International, "HM-1 Integrated Vehicle Health Management Committee," https://www.sae.org/works/committeeHome.do?comtID=TEAHM1.
14. Wikipedia Contributors, "ACARS," https://en.wikipedia.org/wiki/ACARS, accessed December 2017.
15. Kaiser, J. and Vernaleken, C., "Civil Aviation," in *Information Ergonomics: A Theoretical Approach and Practical Experience in Transportation*, Stein, M. and Sandl, P., eds., (Berlin: Springer-Verlag, 2012).
16. Bense, W., "Prognosis and Health Monitoring Systems for Aircraft Engines," SAE Technical Paper 2013-01-2146, 2013, doi:10.4271/2013-01-2146.
17. MIMOSA, "OSA-CBM 3.3.1," http://www.mimosa.org/specs/osa-cbm-331, June 2010, retrieved January 2018.
18. Simon, D.L., "Propulsion Diagnostic Method Evaluation Strategy (Pro-DiMES) User's Guide," NASA Technical Memo TM—2010-215840, January 2010.
19. Federal Aviation Administration, "Electronic Code of Federal Regulations," https://www.ecfr.gov, retrieved December 2017.
20. Rohy, D.A. and Compton, W.A., "Radiation Pyrometer for Gas Turbine Blades," Final Report, NASA Marshall Space Flight Center, October 1973, https://ntrs.nasa.gov/archive/nasa/casi.ntrs.nasa.gov/19740020818.pdf, retrieved January 2018.
21. Trimble, S., "Pratt & Whitney Selects Canadian Supplier for F135," *FlightGlobal*, May 2009. https://www.flightglobal.com/news/articles/pratt-whitney-selects-canadian-supplier-for-f135-327055/, retrieved December 2017.

22. Hess, A., "The Prognostic Requirement for Advanced Sensors and Non-Traditional Detection Technologies," *DARPA/DSO Prognosis Bidder's Conference*, Alexandria, VA, September 2002. http://www.dtic.mil/dtic/tr/fulltext/u2/a513139.pdf, retrieved December 2017.

23. Higgins, P.D., Crow, J.T., "Advances in Commercial Engine Lube Debris Monitoring," SAE Technical Paper 972603, 1997, doi:10.4271/972603.

24. Muir, D. and Howe, B., "In-Line Oil Debris Monitor (ODM) for the Advanced Tactical Fighter Engine," *Technology Showcase: Integrated Monitoring Diagnostics and Prevention, Proceedings of the Joint Conference*, Mobile, AL, April 1996.

25. gastops Inc., "Bearing and Gear Condition Indication," http://www.gastops.com/solutions/aviation/bearing-and-gear-condition-indication/engines/, retrieved December 2017.

26. Ma, M., "Establishment and Application of Aero-Engine Lubricating Oil System Model Based on Flight Test Data," *Proceedings of the International Powertrains, Fuels & Lubricants Meeting*, October 2017, doi:10.4271/2017-01-2215.

27. Greaves, R.W. and White, E.R., "An Overview of Airborne Vibration Monitoring (AVM) Systems," SAE Technical Paper 871731, 1987, doi:10.4271/871731.

28. Watson, M.J., Byington, C.S., and Behbahani, A., "Very High Frequency Monitoring System for Engine Gearbox and Generator Health Management," *Proceedings of the AeroTech Congress & Exhibition*, Los Angeles, CA, September 2007, https://doi.org/10.4271/2007-01-3878.

29. Shay, L.A., "Pratt & Whitney Launches 'EngineWise' for MRO," Aviation Week Network, http://aviationweek.com/paris-air-show-2017/pratt-whitney-launches-enginewise-mro, June 2017, retrieved July 2018.

30. Doel, D.L., "TEMPER—A Gas Path Analysis Tool for Commercial Jet Engines," *Proceedings of the ASME Turbine and Aeroengine Congress*, Cologne, Germany, June 1992.

31. RTInsights Team, "How Rolls-Royce Maintains Jet Engines with the IoT," https://www.rtinsights.com/rolls-royce-jet-engine-maintenance-iot/, October 2016, retrieved December 2017.

32. CAMP Systems Inc., "Engine Health Monitoring," https://www.campsystems.com/Engine-Health-Monitoring, retrieved January 2018.

33. Volponi, A., "Gas Turbine Engine Health Management: Past, Present and Future Trends," *Journal of Engineering for Gas Turbines & Power* 136, no. 5 (May 2014).

34. Litt, J.S., Simon, D.L., Garg, S., Guo, T.-H. et al., "A Survey of Intelligent Control and Health Management Technologies for Aircraft Propulsion Systems," NASA Technical Report NASA/TM—2005-213622, May 2005.

35. Radio Technical Commission for Aeronautics, "Guidelines for Communication, Navigation, Surveillance, and Air Traffic Management (CNS/ATM) Systems Software Integrity Assurance," RTCA DO-278A (EUROCAE ED-109A), December 2011.

36. Radio Technical Commission for Aeronautics, "Software Considerations in Airborne Systems and Equipment Certification," DO-178C (EUROCAE ED-12C), December 2011.

37. Wikipedia Contributors, "ETOPS," https://en.wikipedia.org/wiki/ETOPS, retrieved January 2018.

38. Pfoertner, H., "Long-Term Operation and Maintenance of Engine Monitoring Systems - Recommendations Derived from 15 Years of OLMOS Use by the GAF," *RTO AVT Symposium on "Ageing Mechanisms and Control: Part B - Monitoring and Management of Gas Turbine Fleets for Extended Life and Reduced Costs"*, Manchester, October 2001.

39. Harrison, K.J., "Paris 2011: Volvo Aero Launches Tracking System for Saab Gripen's RM12 Engine," *AINOnline*, June 2011, https://www.ainonline.com/aviation-news/2011-06-23/paris-2011-volvo-aero-launches-tracking-system-saab-gripens-rm12-enginer retrieved January 2018.

40. Genot, F., "Cost Reduction and Engine Life Extension through Engine Life Monitoring at SNECMA," *Proceedings RTO A VT Symposium on "Ageing Mechanisms and Control: Part B - Monitoring and Management of Gas Turbine Fleets for Extended Life and Reduced Costs"*, Manchester, October 2001.

41. Narendra, A., Down, J., Mathews, K., Rajamani, R. et al., "Model-Based Detection of Leaks and Blockages in Pipes," *Proceedings of the ASME TurboExpo 2000*, Munich, Germany, May 2000.

42. Volponi, A.J. and Tang, L., "Improved Engine Health Monitoring Using Full Flight Data and Companion Engine Information, *SAE Int. J. Aerosp.*, 9, no. 1 (2016): 91-102, doi:10.4271/2016-01-2024.

43. Behbahani, A.R., Von Moll, A., Zeller, R., Ordo, J., "Aircraft Integration Challenges and Opportunities for Distributed Intelligent Control, Power, Thermal Management, Diagnostic and Prognostic Systems," SAE Technical Paper 2014-01-2161, 2014, doi:10.4271/2014-01-2161.

44. Rajamani, R., Wang, J., and Jeong, K.-Y., "Condition-Based Maintenance for Aircraft Engines," *Proceedings of the ASME TurboExpo 2004*, Vienna, Austria, June 2004.

45. Giampaulo, A., *Gas Turbine Handbook: Principles and Practices*, 3rd ed., (Lilburn, GA: The Fairmont Press, 2006).

CHAPTER 2

Aircraft Engine Maintenance

E. O. Cooper Pacific
Airmotive Corp.

Inasmuch as this paper deals with the maintenance of modern aircraft engines, it is both fitting and proper to first congratulate our manufacturers for the truly remarkable strides which have been accomplished in recent years together with the results achieved. For not only have they provided the needed additional power and advanced reliability to a new high degree, but they have also taken into consideration the manifold problem of the mechanic and have paid a great deal of attention to serviceability in making the various engine units more accessible in order to facilitate service maintenance.

As an introduction to our subject, I would like to quote a brief history covering the recent development of the aircraft engine in order that you may be more familiar with our problem. All references will apply to the radial air-cooled type of engine inasmuch as this is by far the predominating type in use at the present time. To emphasize this development, the following statistics can be quoted:

> The aircraft engine weight per horsepower ratio has decreased on an average of 40% during the last six years. For example, an engine weighing approximately 1000 lb previously developed 575 horsepower whereas today the same engine greatly improved and with only a slight increase in weight is capable of developing a maximum output of 1100 horsepower at takeoff. Accordingly over a period of six years, weight per horsepower has decreased from 1.75 to approximately 1 lb. This factor alone is greatly responsible for the increased airplane performance of which we can be justly proud today.

REPRINT. Paper presented to the Southern California Section of the Society of Automotive Engineers, April 30, 1937. All papers presented at meetings of the Society are the exclusive property of the Society, from which permission to publish this paper, in full or in part, with credit to the author and the Society, can be obtained upon request. The Society is not responsible for statements or opinions advanced in papers or discussions at its meetings.

> The period between major overhauls has been increased approximately 50%. Whereas the former practice was to overhaul engines, between every 350 to 400 h of operation, they now function perfectly for periods ranging from 550 to 600 h depending on the type of operation in which they are engaged.
>
> Maintenance has also progressed to such an extent that it is now possible and practical to check valve clearances only once during the period between overhauls where just a few years ago, it was common practice to perform this operation every 20 to 30 h of operation.

This achievement was only made possible through the cooperation of all concerned. In no other industry is such close cooperation found and to substantiate this statement, the following facts can be given:

Aircraft engine manufacturers invite constructive criticism. They employ service representatives who are always in close contact with the operators in their own particular territory. Just as soon as trouble is experienced or a defect discovered, the information is given to the engine manufacturer, who, in turn, immediately takes the necessary steps to correct the condition and to so advise all operators of the remedy by the issuance of a Service Bulletin.

These Bulletins are numbered consecutively and are forwarded to all operators using the particular type of engine involved. Thus, through the medium of the Service Bulletin, operators and repair depots are kept advised of all changes and developments and so keep in constant touch with the engine manufacturers.

Twice annually, representatives of all of the airlines and the Bureau of Air Commerce attend special meetings of the airline maintenance committee. These committee meetings were inaugurated solely for the purpose of cooperative discussion concerning airline operation problems. At these meetings, existing problems are discussed and brought to the attention of all operators' representatives. Should a condition under discussion warrant a general action, a resolution is passed and forwarded the particular manufacturer demanding immediate attention. The minutes of these meetings are kept strictly confidential and manufacturers' representatives are only admitted when their own product is being discussed. If information is needed by this committee, the representative is called into the session and questioned. Where corrective action on a specific deficiency has been requested at a previous meeting, a report of the progress being made is demanded by the committee. In this manner, the particular manufacturer is made fully cognizant of the deficiency and at the same time is assured of the full cooperation and aid of all concerned with the problem.

Developments follow closely the problems encountered in service as is the case in other industries. To illustrate the results obtained, we can quote a brief outline covering some of the major maintenance problems together with their solutions.

Considerable difficulty had been experienced with bronze exhaust valve seats burning and pitting due to the excessive heat conditions in their particular location. Larger valves were installed, hollow stem sodium-filled valves fitted, and a number of bronze alloy seat materials were experimented with. All of these helped considerably. However, a satisfactory elimination of this prevalent trouble was not accomplished until the introduction of the steel exhaust valve seats and the Stellite-faced valves. This combination has practically eliminated exhaust valve trouble in all cases where a correct carburetor mixture ratio is maintained.

Excessive roughness at certain engine speeds was encountered on engines equipped with a geared propeller drive. This roughness set up an engine vibration which was disastrous to the installation and was particularly ruinous to propellers. In fact, the condition was so critical as to warrant the placarding of aircraft requesting the pilots not to cruise at certain engine speeds. Apparently, a resonant vibration was being set up within the engine, and in order to finally overcome the same, the crankshaft dynamic dampener was developed and installed. This device consists of a sliding weight with a free circumferential movement of approximately 1/2″ and is installed as a part of the crankshaft rear counterweight. The action of this unit can best be described with reference to a pendulum acting upon a second pendulum in which case the oscillating force of the former is destroyed, thus eliminating the previous inherent vibration. This development alone is largely responsible for the reduction in maintenance problems of the higher output engines and today is a standard installation in all engines where the same are geared or direct drive.

Master rod bearing failures due presumably to lack of oil when the engine was first being started have been eliminated in one instance by the installation of a compensating oil pressure relief valve. This additional valve preloads the engine oil pressure relief valve assembly to provide an increased pressure of approximately 180 lb/in.^{2} when the engine is operated immediately after starting. As soon as the oil has completely circulated through the engine and a pressure of 80 lb/in.^{2} has been obtained, the compensating valve is automatically released. In this manner, a supply of oil is almost immediately assured to all of the vital parts of the engine.

Automatic force-feed circulating oiling systems constantly lubricating the valve mechanism have eliminated the removal of push rods and the consequent checking of valve clearances. This development has also increased valve guide life approximately 100% in addition to providing a needed cooling medium for the valves themselves and the cylinder heads.

Ignition cable was a source of worry to all operators for a considerable period and not until the synthetic rubber composition covering was introduced was this trouble remedied. The earlier types of cable were subject to failures due to their becoming saturated with engine oil, thus destroying their insulating qualities. The synthetic composition oil resisting cables are accordingly rendering an entirely satisfactory service.

Magneto output was questionable at one time. Apparently, the engine manufacturers were making demands far in advance of the available product, with the result that the magneto performance on the higher output engines had to be improved immediately. This emergency was met and a remedy provided primarily through the installation of the cobalt iron rotating magnet and the pivotless breaker assembly which greatly increased the magneto efficiency.

Cylinder wall wear has long been a vexatious problem and regrinding was often necessitated at the time of each major engine overhaul. Today, however, such is not the case. The use of tapered cylinder walls and Nitralloy cylinder sleeves has prolonged this life to as high as 2000 operating hours before a regrind becomes necessary. The tapered cylinder wall provides a taper at the upper end which converges to a −0.010″. Thus, as the cylinder wears, the walls become parallel. This development as stated above has produced very satisfactory results.

Previously, an engine was operated on the ground for long periods in an effort to warm the engine oil preparatory to takeoff, with consequent damage to the engine due to overheating the air-cooled cylinder heads. Today, an automatic oil temperature regulator now permits almost instantaneous takeoff. This procedure is made possible through the use of a sylphon controlled valve which permits the circulation of only a small part of the oil supply until the desired temperature is reached, following which the valve opens permitting the circulation of the entire supply.

However, with all of the aforementioned improvements in service, we are still confronted with a number of unsolved problems which today are claiming the constant attention of the manufacturers' engineers. In order to improve the engine performance now being obtained, the following conditions will necessarily have to be corrected:

Formation of ice within the carburetor with resultant damage to the supercharger impeller.

Entrance of dust and other foreign matter into the carburetor air scoop causing abnormal wear to the cylinder walls and piston rings.

Drainage of oil into the engine due to the gravity feed installation of the oil tank when the aircraft is left standing for a period of several days without being operated. Where this occurs, oil will accumulate in the lower cylinders and intake manifolds and unless extreme precautions are taken to completely drain this oil, there is a possibility of serious damage being caused when the engine is first started inasmuch as same cannot be displaced rapidly.

In many engines, the supercharger impeller nut lock is not adequately secured. Due to the high rotational speed of this assembly which in many cases exceeds 25,000 RPM, a tremendous force is exerted on the lock and as a result a number of failures have occurred. As a suggestion, a lock should be designed, a feature of which would be that the centrifugal force be applied to improve the locking qualities. In this case, the lock would be more positive at the higher operating speeds and during periods of acceleration.

At the present time, the aircraft spark plug requires attention prior to any other unit assembled in the engine. In airline operation, spark plugs are serviced every 65 h whereas the engine does not require a complete service check but once every 125 operating hours. The troubles experienced are chiefly rapid electrode burning and the breaking down of the mica insulation which is exposed within the cylinder. Considerable improvement of this unit would be welcome, although we realize that everything possible is being done to produce the desired results. Leaner air fuel ratios and the use of leaded fuels have greatly affected spark plug life and have consequently created a major problem.

Sticking piston rings are still causing a great deal of worry to the operator although, in most cases, this condition is directly traceable to lean mixture ratios or insufficient lubrication. Ring clearances have been increased in an endeavor to help eliminate this trouble. However, experience has shown that where engines are being operated with the installation of an automatic mixture control carburetor or that an exhaust gas analyzer is used to maintain a constant air fuel ratio, trouble from this source has been reduced to a minimum. In an effort to provide more adequate lubrication to the pistons, master rod bearing clearances have been increased which allow an additional supply of oil to be distributed to the engine power section. However, as previously stated, this condition can still be considered a problem.

In aircraft engine installations, no provisions are made to clean the oil before the same enters the engine from the supply tank with the result that if foreign matter is present in the oil, the same can cause serious damage to the internal parts and bearing surfaces of the engine. As pointed out earlier in the paper, an instantaneous supply of oil to the engine is required at the moment of starting, therefore, the problem of cleaning the oil under those conditions would appear to be unsurmountable. However, due to the number of failures which have been caused by this actually happening, the same is worthy of serious consideration.

Engine operating life between overhauls, unfortunately, is now governed by the oil sludge deposits which accumulate and eventually close the internal lubricating passages. Where it not for this fact, the engine life between overhauls could be greatly extended. At the present time, engines are completely overhauled every 500 to 600 operating hours and it is reasonable to presume, judging from the amount of wear found, that if oil sludge could be entirely eliminated from the engine, the operating period could be extended an additional 200 h.

Inspection is a vital function of engine maintenance and overhaul inasmuch as the increased output of the present-day engine demands that the same be exercised with considerably more precision than was previously required. Aircraft engines are given a visual inspection daily during operation, and a constant record of the performance obtained during flight is given to the operating personnel. Should any irregular condition be reported, an immediate investigation is made and the necessary corrective measures taken before the airplane is returned to service. In this respect, the pilot is kept well advised regarding engine performance being obtained, inasmuch as the regular flight instruments include a cylinder head thermocouple temperature gauge, a manifold pressure gauge, oil pressure and temperature gauges, tachometer, and a positive means of determining mixture fuel-air ratio.

As mentioned previously, the spark plugs are serviced every 65 h. At intervals of 125 h, a more rigid and thorough inspection is given and a specially prepared maintenance form outlining all points to be checked must be filled in and signed by a Government-licensed mechanic who is held responsible for the work completed.

This check consists of a complete examination of the entire engine and accessory installation and includes a general tightening of all parts which might tend to become loose during operation.

The oil is also drained and all screens and filters are carefully cleaned.

In addition to this work, the entire ignition system is given a careful inspection and adjusted where necessary.

Immediately following all such engine checks, a complete record of the same is made in the airplane logbook in accordance with Bureau of Air Commerce regulations.

At the expiration of the 500 to 600 h operating period, the engine is removed from the airplane and given to the repair depot for complete reconditioning.

In the past, it was customary to top overhaul engines between major overhauls, in which case, new piston rings were fitted and valves were ground. However, this procedure is not recommended at the present time and has been abandoned by the majority of operators. Not only was the operation costly, but it has since been proven unnecessary.

The first step of the major overhaul operation is the complete disassembly of the engine. Following this, all parts are thoroughly cleaned and all paint is removed to facilitate the rigid inspection which is to follow.

Prior to the reconditioning being commenced, all engine parts are given a thorough inspection. Dimensions and tolerances are carefully checked and noted by the inspector on the specially prepared overhaul inspection chart which is provided. All defective parts are replaced immediately following this inspection.

In addition to the visual method, all steel parts are inspected by means of the Magnaflux magnetic system. When this is done, each part is magnetized for the purpose of detecting cracks when the same is immersed in a tank containing an agitated solution of light oil and iron powder. Should the part being inspected contain a crack, same is immediately detected due to the presence of the iron particles. Following this inspection, each part is carefully de-magnetized in order to prevent the possibility of metal filings adhering to the part during overhaul.

Due to the complexities of the modern aircraft engine, the overhaul is subdivided and the parts distributed to the various departments. In other words, the mechanics are trained to specialize in one certain branch of overhaul. That is, the power section, the rear section, the cylinder assemblies and the electrical equipment are each given to respective departments for overhaul. Following the reconditioning of each of these subassemblies, the same are returned to the final assembly department, where the engine is completely assembled and made ready for test.

Each engine is then installed on the test stand and is given a 10 h run prior to installation in the airplane. During this test, all functions of the engine are carefully checked and all readings recorded. Following completion of this test, the engine is ready for installation and immediate operation.

The amount of replacement parts installed during overhaul has decreased proportionately with the many improvements which have been incorporated within the engine. Whereas overhaul formerly cost approximately $3.00 per operating hour, the same has been reduced to approximately $1.50 per operating hour at the present time. This figure includes both labor and material. The average complete engine overhaul requires in the neighborhood of 150 man-hours for single-row radial engines. This figure, however, does not include the removal and installation of the engine in the airplane.

Maintenance costs vary greatly according to the type of operation and the weather conditions involved. Therefore, it is almost an impossibility to quote a general cost figure per hour of operation. Needless to say, however, the same has been greatly decreased in terms of dollars per horsepower obtained within the last few years.

Forced landings due to mechanical failures have also been reduced to a minimum, and although a percentage figure is not available, statistics show the same to be almost negligible, which fact in itself speaks well for the engines and the maintenance which they receive.

CHAPTER 3

Aircraft Gas Turbine Engine Trend Analysis

G. H. Weitz and E. J. Briggs Jr.
Federal Aviation Agency

R. S. Stahr
American Airlines, Inc.

J. J. Eden
Trans-Canada Air Lines

W. J. Cake
Pratt & Whitney Aircraft, Div., United Aircraft Corp.

C. A. Fisher
Trans World Airlines, Inc.

FAA Reliability Concept

THE FEDERAL AVIATION ACT of 1958 says, among other things, that: "The administrator shall exercise and perform his powers and duties under this Act in such manner as will best tend to reduce or eliminate the possibility of, or recurrence of, accidents in air transportation." Simply put, this means that the FAA and the industry must see to it that each airplane is in proper condition to fly safely from A to B each and every time. If any difficulty arises which affects safety, that difficulty must be reduced or eliminated.

Our part of the effort concerning safety deals with continuing airworthiness. There are other people in the FAA who concern themselves with air traffic control, airports, air routes and the like.

Considerations

In looking at our part of the job, we are convinced that there are better ways for us to gauge the level of airworthiness being achieved on aircraft in use in air carrier air transportation.

When a new airplane is being judged for compliance with the regulations for type certification purposes, both the manufacturer and the Federal Aviation Agency are able to compare its compliance with a fairly well defined set of standards. Such standards cover the strength requirements for the airplane, its flight performance requirements and its overall configuration requirements.

The airplane, its engines and components, undergoes exhaustive dynamic, static, fatigue and flight tests before acceptance. We well realize the controversies which must be solved before final acceptance by the FAA. Yet, once the airplane is placed into service, our requirements do not reach into the future except to say that the airplane must be adequately maintained so as to continue to be "airworthy."

The term "airworthy" thus immediately becomes controversial and subject to the interpretation of many persons under varying circumstances. The airplane is no longer new. It is now worn to some unknown degree and perhaps prone to some unknown malfunction. We must know, for sure, how reliable it is. We must, therefore, have a measure of the ability of the airplane to fly safely and reliably.

Reliability Yardstick

In examining the meaning of the term "reliability," we have found it to be a relative term dependent upon the state-of-the-art. Some airplanes and engines are better than others because of better integrity of design, overstrength of materials, simplicity, maintainability, or other factors determined by the builder. All of these products meet the minimum standards, of course, but even though good maintenance practices are applied to all, some may not be as reliable as the others since they were not favored by the state-of-the-art.

It is our intention to find out which airplanes are the more favored ones and which are not. We think we can research service experience and define reliability programs using parameters which have some degree of uniformity. When we are able to find such parameters, we will be able to analyze each type airplane–along with its systems and components–and focus our efforts on the weak spots. Simultaneously, once a reliability index (or alert value) has been established for a given type airplane, or its systems or its components, we will be able to judge whether the air carriers are able to meet that value. Additionally, we will be able to tell which air carriers are achieving a better rate and also identify those which consistently are worse. These parameters will be our "reliability yardsticks."

Propulsion System Reliability Program

This program which has been in effect for about two years has given us an opportunity to evaluate, among other things, methods for data collection and use of this data to determine trends.

This program uses the rate of powerplant shutdowns in flight to measure reliability and provides for corrective action should the rate become abnormally high. It places responsibility for initiation of such action upon the carriers and provides both the carriers and the Agency with a means for constant surveillance of the effects of this action.

Overall results indicate that the program has been successful with participation now including about 1500 air carrier aircraft out of some 2000. Under this program we have been able to recognize the inherent high reliability of the turbojet powerplant. This has resulted in a rapid increase in authorized time between overhaul without adversely affecting continuous airworthiness. One model engine is being operated 5000 h between overhauls. Analysis of the data collected has enabled the FAA and industry to spot trouble areas such as fire-warning systems and to provide fixes without penalizing the rest of the engine.

Aircraft Systems and Component Reliability

We are now in process of evaluating other reliability programs dealing with the continued airworthiness of systems and components. This program started in October 1962 and has progressed steadily to the point where we are almost ready to issue national policy material. This policy material will prescribe the types of reliability programs acceptable to us and which can be used in lieu of control methods utilizing hourly overhaul intervals or hard times.

The derivation of the policy has been based on actual study programs that have been going since February 1963. United Air Lines, American Airlines, Trans World Airlines, Pan American, and Eastern Air Lines are participants in these studies. Each of these airlines has been working on reliability programs which are different in concept. For example, United maintains a level of reliability by the use of a sampling program that requires the removal and overhaul of the highest time components each month, and by monitoring the performance of these components using probability of survival and conditional probability of failure techniques. Pan American selected the system approach to reliability. They chose several aircraft systems and from considerable past experience arrived at a calculated failure rate expressed in rate per 1000 h. This required a definition of the term failure or shutdown so that a precise point could be established. For instance, PAA defines a failure of the DC-8 hydraulic power system as one where any reported loss of system pressure or loss of fluid during any A to B flight requires the use of the auxiliary system. Similarly, EAL and TWA have programs in existence which provide reliability parameters using other methods.

Although we are reasonably content with our progress so far, we are not satisfied. As with any program which seems to have some future thinking in it, the future soon becomes the present and then passes quickly into history. We still have accidents, we still have A to B difficulties and under the FAA Act we must continue to perform in such a manner as will best tend to reduce or eliminate the possibility of, or recurrence of such happenings in air transportation.

Perhaps today's plans for the future will give us better maintenance tools to use such as the maintenance recorder. Incidentally, we are installing one in our Convair 880. It is similar to TWA's. Or perhaps we will find better means to internally examine and forecast the life limits of spars, beams, chords, or stabilizer attach points in airframe primary structure. We think this is our next job.

A NUMBER of trends are developing in the field of gas turbine design. Most of these trends are directed toward the goals of greater economy; they are thus inter-related. Briefly these design trends may be listed as follows:

- Design Trend 1: Longer Part Life.
- Design Trend 2: Reducing Frequency of Maintenance Activity
- Design Trend 3: More Systematic Methods of Determining Engine Condition (a) on the Airplane and (b) in the Shop.
- Design Trend 4: Increasing Sophistication in the Use of Statistics.
- Design Trend 5: Higher Standards of Dependability.
- Design Trend 6: Higher Unit Productivity.
- Design Trend 7: Higher First Cost.

With the exception of Design Trend 7, Higher First Cost, all of these design trends are complementary; they are aimed at reducing both the cost and the uncertainties of airline operation and thus increasing its attractiveness to the customer. Trend 7 is included in the list because it is one of the prime incentives for all the other trends.

All of the above trends must be considered in planning the program for maintaining a new engine. One of the key facets of the maintenance program is trend analysis.

The objectives of trend analysis are:

a. to permit uninterrupted operation of sound equipment;
b. to schedule maintenance activity only as needed;
c. to reduce the frequency of unscheduled maintenance actions;
d. to reduce fuel consumption due to performance deterioration; and
e. to reduce expenditures for new parts as a consequence of serious failure.

Let us examine how each of the Design Trends relates to the objectives of trend analysis.

Design Trend 1: Longer Part Life. In general the engine designer is raising his sights in the area of durability and reliability. This trend reflects an awareness of the logarithmic nature of part life. Having found that the jump from 5000 to 10,000 h is little more taxing than the earlier development steps from 1000 to 2000 and then from 2000 to 4000 h, now the designer is looking quite seriously at 20,000 h and higher lives for a high percentage of his engine's parts.

This trend accentuates the need for effective means of monitoring engine health in situ, since the engine will be off the airplane and in the shop less frequently for maintenance action.

Design Trend 2: Reduced Frequency of Maintenance Activity is a direct result of the development long-lived parts coupled with the desire for higher standards of dependability.

It is generally true that a machine that makes infrequent scheduled visits to the shop has less exposure to assembly discrepancies than the machine that requires more frequent attention. For this reason, experience has shown that the overall rate of engine failure usually decreases when TBO is increased.

This trend will continue so long as higher quality and longer life parts are incorporated in new or overhauled engines. However, we must be careful to credit such economic benefits as may result from increases in TBO to the proper accounts.

Design Trend 3: More Systematic Methods of determining engine condition are under development. For in-service use, some methods are aimed primarily at mechanical condition (e.g. magnetic sump plugs, acoustic analyzers, bearing vibration and temperature sensors); others are aimed more at efficiency (e.g. performance parameter recorders).

In the overhaul and repair shop more analytical procedures show some promise of reducing the guesswork associated with overhaul-vs.-repair decisions. Establishing trends in the physical condition of engine hardware will require greater sophistication in the use of sampling techniques and probability theory (Design Trend 4) because fewer engines will be seen in any given time period due to Design Trends 1 and 2.

Current experience indicates that deterioration affecting performance is categorically low for gas turbine engines, and that it will be further reduced with longer life parts. This suggests that elaborate systems for monitoring component performance in routine flight may be expected to have less utility on future power-plants than has been generally accepted.

Design Trend 4: Increasing Sophistication in Use of Statistics is being exhibited by the leading airlines. Trends in failure rates of individual parts and components, as well as the total engine, will receive at least monthly review. Also, the quantitative analysis of wear rates and patterns on those parts that have not actually failed will most certainly be developed further.

This trend tends to reduce the dependability upon systems for measuring engine condition directly in flight—except perhaps in certain cases where the statistical analyses suggest that individual engines bear especially close watching. History shows that most fleets possess at least a few chronic trouble-makers.

Design Trend 5: Higher Standards of Dependability are naturally demanded of a healthy maturing transportation system. The customer using the system, together with competition, sees to this. Analysis of trends that may affect operational reliability gets a big boost from these demands.

Design Trend 6: Higher Unit Productivity, either due to higher aircraft speed and/or higher payload, increases the incentive to anticipate problems before they become serious. Analysis of the several trends in (a) shop inspections, (b) statistical probability of failure, and (c) engine performance before and after overhaul, together present considerable evidence of which engines bear the closest watching. Scheduling of corrective action can be further augmented by the application in flight Early Failure Detection devices and systems.

Design Trend 7: Higher Unit Cost is additive to the above incentive (Design Trend 6). The SST engines, costing over $1 Million each, are an extreme example. But, even the light-twin subsonic jets have powerplants, when fully assembled, costing over $1/3 Million each.

Conclusion: It is clear to this observer that keeping on top of the trends reviewed above is the prime responsibility of airline powerplant engineers. It is not clear exactly how the emphasis on some of the techniques and procedures mentioned will change, but that such emphasis will change is certain. Skillful and alert management will be required to keep pace with the changes.

Exchanges of opinion and judgment, such as that taking place in this panel discussion, are a definite aid to the development of positions for management. It has been both a pleasure and a privilege to be asked to participate.

Gas Turbine Trend Analysis

IN THE OPERATION of Gas Turbine Engines large savings and substantially improved reliability can be achieved by Effective Trend Analysis.

Effective Trend Analysis requires a Total Approach and a Scientific Approach.

Total Approach

Yardsticks have to be established to measure the behaviour of the respective material, as outlined in Figure 1.

Is there a significant change in Properties?

Is there a significant change in Shape?

To answer these two fundamental questions a well defined set of standards and observations have to be established. The standards must be continuously maintained for accuracy as data is obtained and the data has to be continuously examined for significance.

The data system must relate to all significant components of the system. Individual parts, sub-assemblies, assemblies, units, systems and finally the aircraft–each can produce a related or unrelated trend in its own right, which if observed in time can permit required preventive action.

Each of the above components can be monitored by different observations at different stages in its life cycle. A part exhibiting wear or cracking during shop inspection is defining its optimum usage by the evidence it carries.

In addition to the individual detail analysis of the mechanism of attrition, significance of the trend will depend on the relationship to age as well as the detail of crack origination and location or wear pattern as observed on a sample number.

FIGURE 1 The total approach to effective trend analysis.

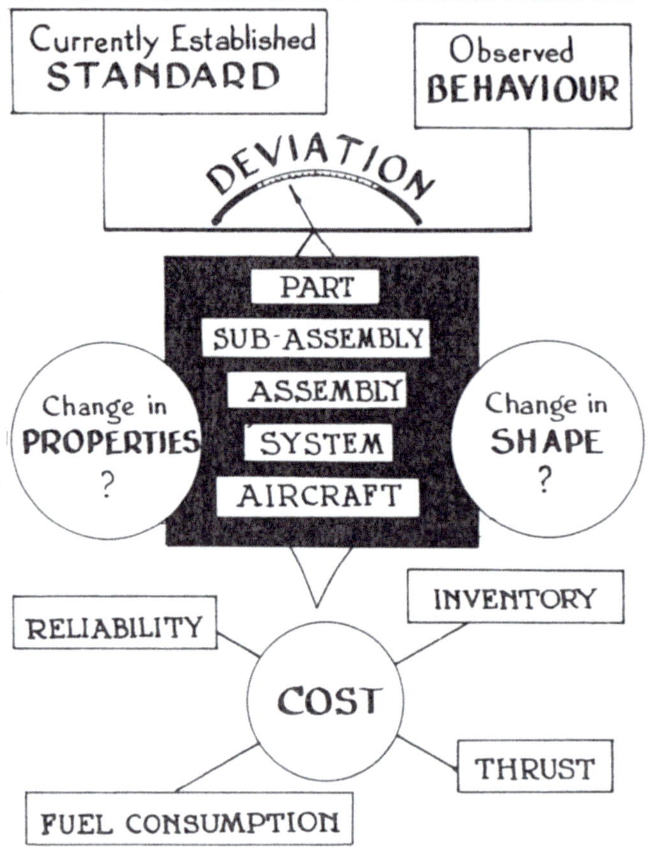

A unit when checked on rig test, or an engine on a test cell, can show certain individual deficiencies or hide these unless examined in the light of overall test results and trend analysis gained from the testing of a number of units.

Maintenance ground checks of equipment require analysis, both individually and in the light of total experience, as do observations in flight. As shown in Figure 2,

the shop
the rig test
the test bed
the aircraft on the ground
the aircraft in flight

are all elements in the life cycle of the:

part
sub-assembly
assembly
system
and aircraft

The data programme has to embrace all these constituents in its Total Approach to provide Effective Trend Analysis.

FIGURE 2 Sources of trend analysis data.

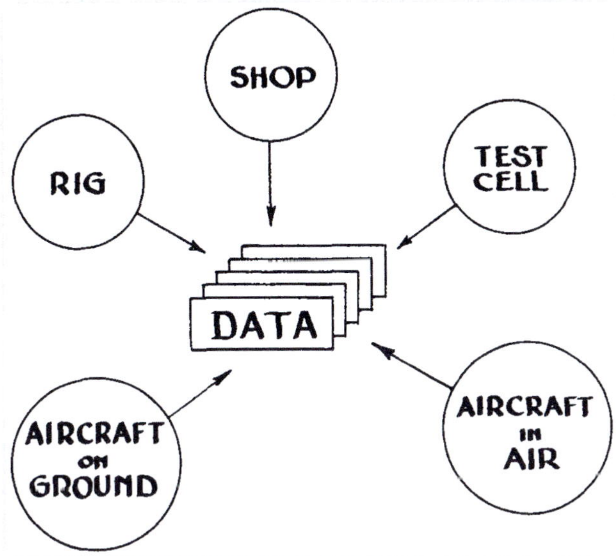

Scientific Approach

Over the past twenty years the application of mathematical and statistical methods to the selection of sample size, confidence in test results, the assessment of significance in observations, etc. has rapidly increased. The use of statistics in Reliability analysis is also gaining ground in the Aviation Industry. A more rigorous analysis of the application of scientific methods to Aviation problems is required. The world of Equipment usage is fertile ground for an injection of practical, comprehensible, down-to-earth statistics-exemplified by application to practical day-to-day problems.

To adequately monitor individual part behaviour, increased usage of risk of failure per time zone is required. Specified data has to be organized and then presented in a fashion so that a simple curve of failure risk can be presented to permit a quick assessment of the trend of behaviour.

For good Trend Analysis the following basic questions have to be answered:

At what total time is this component expected to fail?

At what total time was this component scrapped?

What is the total time at which this component failed or would have failed for this cause?

What is the experienced component population travel through each time zone?

A specified schedule of rig testing is not adequate for unit control. The manufacturer should devote time and talent to the analysis of the significance of test results, both from one unit and a number of units, to help define the data system by which significant trends can be established.

The test cell is probably the best monitor of trend of fuel consumption and thrust with component life and modification standard. A clear definition of required sample size and frequency of after-service test to give significant trend analysis of these important and expensive variables, should be part of the standard overhaul manual.

Preventive maintenance throughout the life cycle of the components used requires knowledgable and systematic analysis. A scientific approach to the definition of the means of observation, the data collection, presentation and analysis is urgently needed. Arbitrary requirements and recommendations in respect to check intervals and methods of checking bring about arbitrary and expensive limitations.

Designs which have not been assessed from the viewpoint of Early Failure Detection introduce the potential of unnecessary shutdowns and failure costs. The difference between a failure detected early and one allowed to develop, can be the difference between a harmless planned removal and a critical inflight failure; as well as the difference between \$500 in material scrappage as compared to \$50,000.

A complete and detailed Gas Turbine Failure Analysis is fundamental to a knowledgable approach to Trend Analysis. The Failure Analysis should categorize the effect of failure and bracket the probability of occurrence. The Failure Analysis is the basic preamble to a detailed analysis of potential Early Failure Detection. This must explore all available means of detection and compare the resultant probability of detection afforded by each of the analyzed methods.

The degree of requirement of techniques such as vibration monitoring, breather pressure checking, streak detection, filter checks, metal detecting plugs and automatic flight performance recording should be defined as an analytical and logical outcome of systematic analysis.

To illustrate the point, take the problem of determining which gas turbine parameters would be usefully recorded for purposes of preventive maintenance. I suggest that the instinctive arbitrary decision to record all those parameters, presented to the pilot for the purposes of flying the aircraft, would not be the most useful parameters to record for preventive maintenance. Only a detail listing of potential failure modes, coupled with an Early Failure Detection analysis, can reveal the best answer.

Early Failure Detection as an afterthought is a sign of lack of recognition of the operational aspects of Equipment Usage. Early Failure Detection has to be an inherent and important part of the total system design considerations, both in respect to components and the failure detection system itself.

The probability of operational failure is not only dependent on the inherent reliability of the component, but with the same degree of significance to the umbrella of Early Failure Detection, as expressed by the formula:

$$P_{of} = P_f\left(1 - P_{ed}\right) \tag{1}$$

where

$P_{of\_}$ is the Probability of Operational failure
P_f is the Probability of Component failure
P_{ed} is the Probability of Early Detection

A detailed analysis of each failure, weighed against the means of detection, will readily pinpoint those failures which do not give adequate warning until it is too late. Such recognition is the key to the required action, since clearly a redesign of the components or of the detection system will be required. Where this cannot be achieved, economically or due to the state of the design art, arbitrary life limitations will have to be imposed.

Data Sharing

A large amount of data is required for good trend analysis. Where each Airline is restricted to information from its own operation, the risk of unobserved out-of-tolerance is larger than where it is part of a larger pool of experience.

In practice a concerted effort is needed to bring us to effectively share the data which is available. A sensible approach to the collection and distribution of data, on the same equipment used in different Airlines, is an essential step towards Effective Trend Analysis. Time allocation and knowledgable effort is needed to make this obvious benefit a practical reality.

Action

Effective Trend Analysis requires effective action. Good trend analysis with poor action is like a fire alarm without a fire brigade.

The yardsticks of Safety
Cost
and Operational Effectiveness

do not sense individual constituents but only mirror the efficiency of the weakest links.

Effective Trend Analysis, therefore needs alert and continuous attention, pin pointing the weak links and exposing potential areas of improvement to an objective review of investments against returns.

In my opinion, the leverage at our present stage of development is such that the returns in providing a more total and more scientific approach to Gas Turbine Trend Analysis will far outweigh the investment.

Conclusions

Sufficient experience is now available to show the great significance of Effective Trend Analysis in Gas Turbine Engine operation.

An injection of practical statistics, applicable to problems of aircraft Equipment Usage, would be of direct and immediate benefit.

The SAE publication "Reliability Control in Aerospace Equipment Development" sets a useful pattern. A similar publication on "Reliability Control in Aerospace Equipment Usage" would find a warm welcome in the Airline Industry.

The concept of Trend Analysis and Early Failure Detection requires recognition as a basic part of the design programme and cannot be left as an afterthought.

The methods, intervals, techniques and equipment required for Effective Trend Analysis should not be arbitrarily selected as in the past, but should be a logical outcome of a detailed Failure and Early Failure Detection analysis.

The organization of data to provide a means of information interchange between operators of the same equipment, is fundamental to improving the economics of any single operation. Knowledgeable and concentrated effort by the Airlines is required to make this a practical reality.

Total and Scientific Trend Analysis is fundamental to Safe, Economic and Reliable Gas Turbine Operation.

THE OPERATORS OF JET-powered aircraft have been looking for ways to monitor their engines' condition since these aircraft began flying around 1945. Many processes have been tried-oil analysis, run-down time, oil consumption, vibration monitoring, time-temperature recording and others. All of these are useful, but no one by itself has proven adequate for complete monitoring, although each is still being used in one application or another.

When the United States airlines began commercial jet operation in 1958, a significant new set of parameters was added to the engine monitoring picture. Pratt & Whitney

Aircraft had been the only engine manufacturer to insist on the use of engine pressure ratio (EPR) to set thrust in the earlier military aircraft, and had therefore provided as well the unique capability, which EPR allows, of monitoring engine condition. A gas turbine engine with a fixed area exhaust nozzle and good exhaust gas pressure sampling probes will follow its corrected gas generator calibration curves more faithfully and more precisely than most of our present instruments can follow. Gas generator analysis had been used as an engineering tool for many years, but it remained for the commercial operators to make a good service monitoring tool of it. Today, almost all of the major airlines in the world use this monitoring method in one form or another, along with the more usual means noted before.

In 1958, the military officially began a long quest for engine monitoring capability by issuing on 23 July their original Operational Support Requirement, OSR-322, for a Jet Engine Analyzer System. The purpose of OSR-322 was "to provide a jet engine analyzer system, consisting of airborne and ground components, to measure jet engine performance, diagnose and isolate troubles, and aid in predicting engine malfunctions and failures." The military people who wrote OSR-322 covered the field much better than they probably realized at the time, in view of the fact that specific equipment to achieve their overall objective did not then exist, and even now is operating only in prototype form.

The time interval involved here is about seven years. What caused this delay? My own opinion is that the time was not ripe in 1958. The several parts of the picture were happening more or less simultaneously-stating the military requirement, bringing the art of electronics up to the point where a practical airborne unit could be made, and getting commercial experience in using gas generator monitoring techniques. The whole idea had to mature a little before industry could produce a practical system. As my contribution to this panel discussion, I would like to outline what our commercial experience has taught us about which parameters to use and how to use them, and to express an opinion on what recording means appear necessary and practical. We believe we now know how to monitor turbine engines, although much remains to be learned about the processes involved.

What Parameters to Measure

A monitoring device which sensed all the parameters read on an engine development test stand obviously could not be carried around and maintained in service aircraft. This is the area that has been solved by using corrected gas generator characteristics, changing the question from "what parameters" to "how many." These parameters-generally RPM, exhaust gas temperature (EGT) and fuel flow, with EPR as the major variable-show both positive and negative evidence. The gas generator is either on its original calibration curves or is off for some good reason. If only a single parameter changes, look for a bad instrument system because it may not be the engine. A real change, either internal or external to the engine, makes more than one parameter move off its basic calibration lines. When one gets into the details of how many engine and aircraft malfunctions will show on these parameters, the list is already large and is growing. Some of these are instrument calibrations, dirty compressors failed turbine blades, exhaust nozzle malfunctions and excess airbleed.

The basic parameters to use for any turbine engine, therefore, are RPM, EGT and fuel flow, with EPR and compressor inlet pressure and temperature sensed to establish the standard gas generator characteristics. If the airborne unit contains calculating and memory functions, the latter three parameters need not be recorded. In addition, a number of other quantities are open to consideration, and experience has shown that the essential and useful ones are easy to determine if it is done for each specific aircraft-engine combination. These are called "mechanical parameters" to differentiate them from the gas generator "performance" parameters, and consist of such things as oil pressure, breather pressure, vibration, etc. Actually, we feel that every turbine powered aircraft should have variation instrumentation aboard at all times, but there are some problems involved in its installation and use.

The simplest and most effective way to decide on what mechanical parameters to include is to examine the service record of the specific aircraft model and list the most prevalent serious problems, beginning with the worst. The listing can be continued as long as the operator will agree to buy and fly the instruments. There is a practical limit, of course, and one monitoring device now flying records N_2 rpm, EGT, fuel flow, oil pressure and vibration level.

In considering what parameters to measure, one must also decide whether he is trying to monitor only the machinery or whether he is monitoring the pilot's use of the machinery as well. While the operators may have to decide on this point, the engine manufacturer, fortunately, does not get involved in this interesting question except in an advisory capacity.

Recording the Data

How should the data be recorded? A flight recorder is many things to many people, but they can be roughly divided into three classes. These would be maintenance recorders, crash recorders and flight test recorders. There appears to be no satisfactory way to combine these uses into one recorder as their basic purposes are too different. Consequently, it seems apparent that the aircraft which carry crash recorders also will need a separate maintenance recorder, since the crash-proof case and the type of tape record make the crash recorder difficult to use for maintenance purposes.

Several installations of flight test-type recorders have been made and they have produced a lot of in-flight data. However, these recorders have several inherent disadvantages in that their magnetic tape records cannot be read directly by the flight crew, they now can be read out only in ground-based calculators, and it seems doubtful that line maintenance personnel can learn to discriminate quickly amongst all the data collected, to know that the aircraft is ready to proceed on the next leg of its flight. From a human factors viewpoint, it seems necessary to strictly limit the amount of data taken, as well as the number of parameters used, to eliminate as much confusion as possible. If the same brief and clear story can be shown to all people involved-flight crew, ground crew and engineering personnel-they will be able to agree more easily on the meaning of the engine's message.

Monitoring the engines does not require continuously taken data but can better be done by having an airborne device look constantly at the engine's parameters and automatically take data only when a significant limit is exceeded. This must be done without attention from the crew in such an emergency. Other than recording automatically when a limit is exceeded, the unit needs to sample and record data only at selected intervals and its controls should be arranged to do this. These data should be recorded in two forms; direct (uncorrected) data during an emergency or whenever desired in normal operation, and corrected data taken about once each flight to establish gas generator trends. The ideal way to record these data would be on a large, power-house-type paper chart, but a number of considerations enter here. The important points are that the record must be immediately available for use when obtained and the recording must be legible, understandable and a permanent record for use by flight and ground crews, as well as engineering researchers. A large percentage of even the selectively taken data will be useful only in a negative sense in that it will show normal engine operation in all respects.

Using the Data

The commercial airlines, with the assistance of Pratt & Whitney Aircraft, have worked out several systems for monitoring their engines using the data normally recorded in flight by the crew. One system is to compare each engine's data with the average across the wing for each parameter, using the observed readings adjusted for EPR, and perhaps airbleed, differences. When an engine begins to deviate in its relation to the other engines, the crew knows that it must be watched and reported. A second system consists of processing the hand-taken flight log data in ground calculators, correcting it to standard atmospheric conditions and comparing it either to the other engines on the wing or to an absolute standard. This system unfortunately can result in several days lag between the first observation and confirmation of the trend. A third system is to provide the flight engineer with charts or slide rule to correct the data as soon as it is recorded, and plot the corrected data while in flight. While this system is an added duty for the flight crew, it does result in immediate knowledge of engine condition.

An ideal system must provide the answers immediately upon taking the data, and with a degree of accuracy better than is available from hand-taken data. Significant gas generator trends can be as little as 1% or 2%, and the average quality of hand-taken data tends to obscure these shifts in some cases. Data taken and processed electronically appears to be essential. The flight and ground crews must have the best possible data to have a factual basis for judging each situation. This is being done in one airborne device, and the results are very encouraging.

In actually using the data, certain items stand out as being very obvious. An engine overtemperature, for instance, will show up immediately on the recorder and will also cause the flight crew to take some action to relieve the condition. The more long-term, lesser known gas generator trends are harder to find, and require more judgment to interpret. However, this judgment is now being developed and used by the commercial airline flight crews and line maintenance personnel, and it seems essential that these groups continue to exercise this function without undue reliance on professional engineering talent.

Summary

In summary, the aircraft industry needs to agree on answers to the following questions if it is to benefit from the well-proven capability to monitor turbine engines.

1. What parameters must be monitored.
2. Is only the machinery being monitored, or the way the machinery is used as well.
3. How should the data be recorded.
4. How can the data be used.

We at Pratt & Whitney Aircraft believe we have found satisfactory answers to these questions for our equipment. Our efforts over the past few years have been to assist the operators in using these processes to help keep their engines operating, and to interest the instrument manufacturers in making a device to answer the needs we see. I, personally, feel that we can significantly reduce in-flight shutdowns thus providing important economic benefits to the aircraft operators.

Gas Turbine Trend Analysis—Four Phases of Development

DURING THE PAST FEW YEARS, the "trend" has been clearly toward "trend analysis" and it is timely to pause and objectively appraise our progress. What are the different operators doing?, why?, what results?, and what are their plans for the future?

Immediately we are faced with the first area of confusion—different opinions regarding the meaning of trend analysis. To some, it pertains solely to recording and observing the stability of the basic engine performance parameters—EPR, fuel flow, RPM and EGT. To others, it involves monitoring the performance of accessories and components of the power plant systems as well as the basic engine—obviously a much broader scope of activity.

The objectives established by the operators automatically define the scope of their program. These generally include one or more of the following:

1. Provide Line Maintenance and specialist personnel with an additional tool to enhance their ability to detect and analyze malfunctioning engines and accessories, thereby reducing the unscheduled engine removals, inflight shutdowns, delays and cancellations attributable to power plants.
2. Provide a means for obtaining greater utilization of the engines and associated equipment through more effective trouble shooting, extended maintenance and overhaul periods, and possibly on-condition maintenance.
3. Effect a reduction in power plant maintenance and overhaul costs.
4. Improve the over-all operating efficiency of the fleet.

The program at TWA is broad in scope and has developed in four phases, the first two pre-jet and the last two post-jet. The first phase, which evolved out of a program to improve performance of a particularly troublesome piston engine, was basically a flight termination debriefing of the flight crew and analysis of flight log entries (RPM, BMEP, MAP, FF, etc.) by power plant specialists. Any adverse trends indicated by this analysis established the basis for corrective action. Although in retrospect the program was an obvious and simple step, it was actually a major breakthrough into the specialty foreman concept of doing maintenance which has subsequently paid big dividends to TWA and other operators.

The second phase followed closely. In order to provide continuity for the line station activity and a back-up analysis of the flight log data, a team of the most experienced and well qualified specialists was established at the Technical Services Headquarters in Kansas City. Engine operating data in every log sheet was again analyzed for adverse trends and any discrepancy recorded on a large display board for quick visual reference. Where further corrective action was indicated by a pattern on the board, it was initiated without delay through a central maintenance coordinator function, This program satisfied our needs up to the onset of the jet powered airplanes when it became obvious that a faster, more sensitive and efficient means was needed. We could no longer tolerate the 3 to 4 days delay involved in transmitting the flight logs and it became necessary to reduce the data to standard conditions in order to effectively establish trends on individual engines, compare one engine with another, and monitor the over-all fleet performance.

To circumvent these difficulties, a system (the third phase of our development) was devised wherein the basic performance data recorded by the flight engineer is transmitted once daily by teletype to a receiver in Kansas City. Here it is fed into a computer program where it is reduced and compared to previous recordings as well as the design parameters provided by the engine manufacturer. A daily print-out from the machines provides the power plant engineers with a report of those engines having adverse trends or exceeding specified limits. Almost a year has been spent developing this system, primarily in refining the computer program. We believe it is now in the home stretch and will soon be a very effective and valuable tool for maintaining a high standard of engine performance.

The fourth phase, which is still under development, involves the use of an In-Flight Maintenance Recorder to record the basic engine parameters and measurements of other engine accessories. At flight termination, the maintenance specialist will be provided with a complete log of data removed from the recorder for analysis of malfunctions or adverse trends. This system offers the potential of monitoring more parameters than are now available to the flight engineer—or than he could possibly monitor were more instruments made available to him. Secondly, it provides a virtually continuous log analysis—instead of readings taken once per hour or per flight as the case may be—and third, it eliminates the potential errors ever present in transposing instrument readings onto hand-written logs. A Lockheed-Royston, multi-channel, airborne magnetic tape recorder was installed in one 707 airplane in June of last year. Although the usual development problems have been experienced, the unit offers sufficient promise and the over-all objective warrants expanding the program to include a second unit in another airplane which will be installed in the next few months.

Installation of the recorder with its large capacity presented the problem of determining what additional engine and accessory parameters would be most valuable and useful to the power plant specialists. To answer this question, the past several years performance (delays, in-flight shutdowns and unscheduled removals) of each engine model was reviewed. For each malfunctioning, the present method of detection and control and other methods and instrumentation that could be applied or developed for application to the recorder were analyzed. Through this study, we collected data to evaluate against the objectives of a gas turbine trend analysis program.

Consider first the power plant delays of which 90% are caused by accessories and components in the auxiliary systems—fuel and oil, fire warning, start and pneumatic, instrument and control, etc. and only 10% by the basic engine. Measurement of EPR, FF, EGT and RPM alone obviously will not affect the delays materially. A better method is needed for earlier detection of such things as defective fire warning elements, malfunctioning fuel controls, starters, ignitors, incipient fuel and oil leaks and so on to a total of almost fifty different components. In addition, we must be adaptable and flexible because old problems are fixed and new ones develop at a rate of about 50% a year. For a trend analysis program to effect a reduction in delays, it must provide for monitoring the performance of these many components and signalling deterioration in time for preventive maintenance to be taken—no small chore! Trending the basic engine parameters by themselves will unfortunately rarely help that passenger sitting in the terminal waiting for departure of a delayed flight.

Similar to delays, the majority, about 70%, of the in-flight engine shutdowns are caused by the accessories and engine installation components as compared to 30% for the basic engine. For the most part, the components causing in-flight shutdowns are the same as those causing delays, so the additional new instruments required to reduce delays will also reduce shutdowns. However, this relationship is not direct because the majority of failures or malfunctions that cause shutdowns are of the type that develop quickly and instrumentation to detect incipient failure or malfunction before failure is more difficult—a gasket or seal suddenly fails, a switch abruptly malfunctions, or a fuel pump shaft shears, etc. So here again, for a trend analysis program to be entirely effective, it must monitor considerably more than EPR, FF, EGT and RPM.

In one form or another, trend analysis has proved its usefulness in determining whether and when a malfunctioning engine requires removal for repair or overhaul. Many of the malfunctionings are such that if detected early enough, some additional operation of the engine can be permitted. In these cases, the premature removal can actually be scheduled at the most desirable place and time and resultant delays avoided. Through the TWA program, 88% of the removals are made at the few major stations which have spare engines and more extensive maintenance facilities.

It is important to observe, however, that although the trend analysis program unquestionably assists in controlling the unscheduled removals, it does not materially affect the over-all removal rate. It helps to find the fault, but once found, the engine normally requires removal for shop repair or overhaul. Further, the causes of unscheduled removals are ever-changing. Modifications are made to correct specific failures, but new troubles constantly develop. A trend analysis program must therefore be flexible and easily changed to monitor new and different parameters.

There is some thought that an effective trend analysis program will provide a foundation for an On-Condition maintenance program. This is being studied in some depth in conjunction with the In-flight Maintenance Recorder activity. If successful, it will provide a means for increasing the utilization of all components involved and reduce the maintenance and overhaul costs accordingly. For such a program to be workable, the reliability of the recorder and special instrumentation must be well established, which may present a major problem in itself.

It appears that the most promising area for direct savings from a trend analysis program is in the area of engine and aircraft performance. By closely monitoring the specific fuel consumption on each engine, adverse trends can be observed and corrective action promptly established to minimize the over-all costs. It will also provide a means for monitoring the over-all aircraft and engine performance. With fuel and engine overhaul costs totalling upwards to $75 million, only a small improvement is needed to more than offset the costs of a trend analysis program.

Despite the considerable effort of all operators, it appears at this stage of development that many questions regarding the needs and merits of gas turbine trend analysis programs remain to be clearly resolved:

1. What are the specific objectives of a trend analysis program? These may vary considerably between operators depending on each airline's specific requirements and type of operation.
2. What parameters on the basic engine should be monitored?
3. What units or systems in addition to the basic engine should be monitored?
4. What special instrumentation is required?

   What readout or recording frequency,

   What accuracy,

   What availability,

   What reliability, and

   What cost?
5. What other components or systems on the airplane need monitoring and what effect will these have on the power plant requirements?
6. What is the saving or otherwise measurable return to the operator?

The most promising area for direct and tangible saving from a gas turbine trend analysis program is through an improvement of the over-all efficiency of the engines and airplane—specific fuel consumption. In order to effect delay rates, the accessories and power plant systems as well as the basic engine must be monitored. This presents some formidable instrumentation problems yet to be resolved. This need exists on present day airplanes, and will be even greater on the supersonic transport or on smaller two-man crew jet transports where the time available to manually observe and record data will be more limited.

TWA is convinced that trend analysis is an essential part of the maintenance program for all large jet powered transport aircraft. Although many questions remain unanswered, they are concerned primarily with determining the scope of program needed by specific operators. The need for a basic program has been well established and the effectiveness well proven. Developments to expand and improve upon the basic programs must continue with full cooperation and free interchange of information in order that this knowledge can be better applied to the next series of jet powered transports.

Engine Condition Monitoring Techniques for the Future

IMPROVED TECHNIQUES ARE needed for monitoring engine condition while in service. This should be accomplished continuously, accurately, and in depth with the aim of making the prediction and assurance of safe flight more feasible.

The choice of a particular monitoring system is up to the aircraft operator, as the FAA is interested only in the results which are achieved.

In general, operators utilize the currently recognized maintenance practices and systems which, in lieu of systems more specifically responsive to actual need, are based on somewhat arbitrarily established, uniform TBO's and inspection intervals. While promoting and seeking adequate safety, the FAA conducts general surveillance and spot checks of the aircraft operators and of the practices and operational experiences. Accordingly, there is a shared responsibility for promoting safety by both the operators and the FAA.

The dual safety responsibility of FAA and operators has been carried on for many years, achieving generally good levels of safety, however, the desire to reduce costs of maintenance of the more expensive aircraft has brought about a climate of change toward maintenance practices, causing some concern for the future. Specifically, new maintenance control techniques are needed in order for engines to be safely operated over longer TBO's.

Operators wish to control their own TBO's as fully as possible as a means of using their equipment to its fullest safe capacity, and control their costs more effectively. The expansion of air transportation continues, with passenger travel expected to double within the next 10 years.

Turbine engines are generally capable of longer TBO's than piston engines and this has been well demonstrated. Nevertheless, these complex engines have tendencies toward initial failures, wear-out failures, and random types of failures, like engines always have. In addition, there are types of troubles possible in turbine engines, which appear only after many thousands of hours of service, such as creep, time-temperature life limit, and our old friend fatigue. Erosion of aerodynamic shapes and gas flow passages, plus the possibility of high energy rotor burst, complete the picture. The recently granted TBO, of over 5000 h for a turbojet engine, means that these engines may remain in active service for over two years without an opportunity for inspection of all parts. Inspection has been the time honored and the only sure way of determining the existing condition of all engine parts and the basis for predicting their usefulness for further service.

Engine Design Considerations

Basic engine design excellence can insure safe, reliable, and long-life engines, and each engine designer naturally strives to attain the goal of a trouble-free engine. Unfortunately, current design engines are too expensive to be operated for only one run as throw-aways, and even with continuing modification thru engineering support, new troubles seem to appear with extended operating times after the early troubles are solved. If this was not the norm, then the engine would be deemed too heavy, would probably be up rerated, and this would breed a new pattern of troubles.

The Feasibility of TBO Control Based on Condition Monitoring

It is possible that engine condition monitoring may some day be developed to the extent that most of the failure mechanisms are monitored and will provide warning of troubles before they become serious failures. If this is achieved, uniform TBO's and line inspection intervals may possibly be discarded. Such schemes, therefore, would rely on the ultimate principle that each engine "indicates" its condition and when it requires repair or overhaul. An important prerequisite would be that the expected random failures could be tolerated, and that the monitoring system was of better reliability than the engine. Until a system as effective as just related is available, there will have to be somewhat arbitrary but uniform inspection intervals and overhaul periods.

Piston engine problems are still of considerable concern and reserve all the effective controlling action which is reasonable to apply. New data gathering and analyzing techniques should therefore be applied where possible. Ignition and engine analyzer and vibration pickups all with appropriate interpreting techniques have been successful and should be continued. Engine lubricant sample analysis at frequent periods, oil-out temperatures, individual cylinder head temperature indications, and possibly several other engine temperatures and internal pressures might provide worthwhile information on engine "condition." Accurate knowledge of fuel flow, and BMEP have provided one basis for estimating engine operating condition. Operating crew observations are desired for the usual cockpit instrumentation to assist in actual operation and trouble shooting.

Among presently used or proposed engine condition monitoring functions for turbine engines, which are regarded as the most promising, are the performance parameters (r.p.m., gas temperature, hp:, or EPR, fuel flow) and vibration amplitude, time-temperature history of critical hot parts, engine rundown time, sonic analysis, and lubricant analysis. All of these quantities may be automatically recorded inflight except the last three mentioned. Such recorded data should be available for analysis very soon after each flight in order to assess the need for repair or engine change prior to the next scheduled flight. However, until a well proven monitoring system is available, present methods of controlling inspection frequency and TBO will be required.

Condition Monitoring as Related to Engine Failures

The following list indicates typical engine failures which have caused serious malfunctions in service to turbine engines:

Rotor disc and rotor spacer disintegration.

Lubrication system failure.

Fuel & power control malfunction.

Combustor liner distortion, wear, cracking, and burnout.

Fuel manifold cracking or leaking.

Overtemperature operation.

Bearing deterioration and failure.

External flammable line and seal leakage.

Massive rotor blade failures.

FATIGUE - In many of these malfunctions, fatigue cracking has been responsible, particularly in the case of rotor disc, spacer, and blade failures. The possibilities of successfully detecting incipient fatigue damage prior to cracking would appear unachievable at present; however, a breakthrough in these techniques may well be provided. When cracking occurs in a fatigue failure process, usually only an inspection of the part can disclose the crack. Unless this crack progresses very slowly across the metal section there is small hope of inspecting frequently enough to catch all similar cracking in time, in like engines before complete separation occurs. Detection of a crack in a component by means other than visual inspection, is only possible if the crack causes direct changes in engine operation such as engine vibration increase, or internal misalignment or interference. Fatigue of external lines may show up as excess flow, visual leakage, control malfunction or fires. It is often the circumstance that a safe operating period before fatigue crack damage occurs can be established thru reported failure experiences of the part. This often aids in planning parts inspection and/or replacement. Sensitive detectors of vibration, temperature, or rotor turning friction may be useful in evaluating possible initial symptoms of engine parts failing by cracking.

What is fundamentally needed is some way to inspect for and detect the accumulated critical stress cycles which have been imposed on each part. If such data were available, the past stress history of the particular part could be related to its projected future use to determine the remaining safe life. An alternative to detection of stress history would be so-treating all such metal components to nullify the stress cycle grain damage already imposed. In lieu of any of these last "wonder drug" techniques, there must be continued emphasis on engine testing for long time cyclic loading effects, accompanied by designing to low stress levels.

HOT PARTS - Many engine hot part failures are caused by unsuspected overtemperaturing. Faulty thermocouples and lack of automatically recorded temperature histories contribute to this situation. Recording individual EGT thermocouple readings would greatly assist in identifying combustor flame pattern distortions and insure carrying out prompt fixes. Recording the time-temperature history of an engine will provide data which in due course could be related to the safe operating life of many hot parts of that type engine.

VIBRATION INDICATORS - Bearing distress and blade failures have been detected successfully with vibration indicating equipment. Vibration level changes are of prime importance in the use of such equipment. This changing value as compared to an absolute valve makes a rapid decision for corrective action somewhat difficult to make. The FAA has recognized the value such equipment a year ago by allowing 100 h initial TBO increase for turbine engines if acceptable airborne vibration indicating systems are in fleet use. The use of such aids is being adopted widely by many airlines.

LUBRICATION - Bearing wear has been detected through spectrographic analysis of lubricating oil in both turbine and piston engines. This has been used effectively when frequent oil sampling (approx. 20 h intervals) is accomplished. It's application to fleet size programs presents some problems but these should be surmountable. Accurate indicators of internal lubrication and cooling flows or temperatures would be of some value as early warning devices if numerous critical points in the engine were served.

Diametral clearance indicators might prove practical for monitoring rotor disc and blade stretching.

ENGINE FLOW FUNCTIONS - Fuel and power control malfunctions in turbine engines have resulted in engine die-out, unstable operation, overtemperaturing and destructive overspeeding. Critical wear of some internal components and seals with increased TBO's were involved in some instances. It is possible that indicators of internal flows or pressures might have served a useful purpose in these instances. Engine output parameter data might serve to warn of such internal difficulties. Failures of external lines and seals have not been a common fault, but occasionally, severe malfunctioning conditions and fires have been encountered. Perhaps indicators of detailed fluid and air flow trends would be helpful if available.

The addition of numerous temperature, pressure, and flow indicators in engines as suggested here might be rather impractical for various reasons. The use of some of these indication devices at strategic locations as indicated by failure experience is recommended however. An example of this is the oil "out" temperature reading on a particular piston engine some years ago to aid in identifying bearing distress. The usual oil "in" temperature rose so much slower that crew action was too late by that time. Other examples are oil throw-off temperatures, bearing shell temperatures, and externally mounted overheat warning pickups around the engine hot section areas.

TURBINE ENGINE BLADES AND VANES - Rotor blade failures normally result in noticeable engine vibration and often an increase in gas temperature with an accompanying reduction of output. Loss of even single blades may cause added damage by affecting downstream stages and are to be avoided at all costs. Since foreign object damage, creep, root or shroud wear, or excessive temperatures in combinations may be responsible for particular failures, spotting an incipient condition likely to cause failure is difficult without the aid of visual inspection. In situations where centrifugal stretching occurs to a measureable extent before failure, a running clearance indicator may prove useful for turbine engines blades and possibly the rotor discs.

It is hoped that the recent experience gained on the use of the Lockheed and Norden trubine engine data recorders will point the way toward obtaining the "ultimate" in techniques for engine condition monitoring. Use of some advanced condition monitoring techniques for future supersonic transport engines would appear to be virtually mandatory to achieve the goal of economic suitability and feasibility. This view is held because of greater design complexities, the higher cost of "down" times and the probable greater risks of in-flight engine failures.

It is believed that a brief review of typical U.S. fleet engine failure rates will be of interest at this time.

TABLE 1 Average engine reliability data for several airline engines

Year		Engine hours per failure					(1st 6 Mos.)	(1st 9 Mos.)
		1957	1958	1959	1960	1961	1962	1963
Piston engine	A	8000	8000	5500	9000	8000	7000	7700
Piston engine	B	2400	2000	2500	4900	5500	5000	5800
Turboprop	C	14,000	12,000	18,000	36,000	40,000	30,000	80,000
Turboprop	D			9000	2000	9000	10,200	7500
Turbojet	E			32,000	74,000	21,500	41,000	29,000
Turbojet	F				24,000	65,000	22,000	26,000
Turbojet	G				∞	4000	∞	45,700
Turbofan	H					14,000	10,500	40,200

While failure rate is of some significance as an index for evaluating the adequacy of engine maintenance programs, additional factors are in use. Included are (1) the seriousness of each failure, (2) interruptions to output during scheduled operations such as unwanted power reductions, flameout, engine shut-off, propeller feathering, (3) time since inspection, overhaul or new for each interruption, (4) frequency of above events, (5) corrective action programs in progress.

As indicated in the table, engine hours per failure for current highly developed turbine engines have reached outstandingly high levels of well over 30,000 h in some years. In a few instances, as for engines A, D, E and F the level appears to be essentially constant. Engine B, C, and H show improvement over the period shown in Engine G shows consistently high performance.

Several other piston engines not shown which have been used far and wide for many years show times to failure of 2000 to 10,000 h. These engines have not had continued engineering support, hence these levels are expected to represent their peak reliability.

The reliability figures for the current turbine engines are achieved in spite of generally increasing TBO's of from 500 to 1000 h per year for the past 5 years. The TBO's for piston engines appear to have generally leveled off.

References

1. Spraker, W.A., Loomis, J.P., and Begley, "Trend Analysis as Applied to Gas-Turbine Engines and Aircraft Mechanical Equipment," Battelle Memorial Institute.
2. Rolle & Hoekstra, "Propulsion Systems for Civil Aircraft," FAA (SAE Paper).

CHAPTER 4

Maximum Installed Engine Utilization through Disciplined Health Monitoring

Heinz W. Bart and Alois B. Nadler
SWISSAIR Transport Co., Ltd. (Switzerland)

Many airlines find themselves at present in a transition period with regard to engine health monitoring. The manual concepts used up to now are being replaced by computerized versions and almost parallel to this comes the introduction of the "On Condition Modular Maintenance." The new High Bypass Engines (engines of the third generation) along with the second-generation engines were designed for "Modular Maintenance." As a result, the efforts toward achieving an automatic engine performance control were also intensified. These new ideas of engine monitoring were repeatedly considered in determining the Modular Maintenance Concept.

Both in the past and at present, airlines have to some extent applied very different engine trend monitoring methods, in theory and in practice. This shows that even good methods are not universally suitable.

Swissair, which is completely autonomous on the one hand but which works in cooperation with the larger unit of KSSU, has specific problems in the exchange of information with regard to engine health monitoring. We should like to explain the basis of Swissair's various health monitoring methods.

TO KEEP AN ENGINE as long as possible on the aircraft does not necessarily prove an economic operation. It should therefore be explained which fundamental considerations have led to Swissair's somewhat unconventional methods of monitoring.

Although Swissair is fully autonomous, the engine overhauls in KSSU are distributed between the partners. For example, the DC-8 and DC-10 engines (JT3D and CF6-50) are overhauled by KLM, the B-747 JT9D engines by SAS, and finally the JT8D engines of the DC-9 for the whole of KSSU by Swissair. The engineering responsibility for these engines is similarly divided. However, as long as the engine is installed in the aircraft, the health monitoring responsibility lies with the airline concerned. So it is understandable why the engine monitoring is divided into two main groups.

The basis of Swissair's Aircraft Maintenance Concept forms the "Concept of Aircraft Reliability Evaluation" (CARE) which fulfills the requirements of the FAA Advisory Circular 120-17. For engine supervision, the Engine Reliability Program (ERP) was worked out within this Concept.

Since 1972, all of our engines have been maintained solely according to the principles of the "On Condition Modular Concept."

Engine Monitoring Problems

FUNDAMENTAL CONSIDERATIONS—Information which is important for an engine overhaul in the shop is not subject to such pressure time-wise as that which is vital for daily maintenance. The kind of information differs in just the same way. The former primarily describes the general condition, whereas the latter actually leads to localization of the trouble. This time factor and the kind of information often determine the suitability of a certain monitoring concept.

In addition to this, each airline has its own concept regarding the use of its material and equipment. Those of Swissair influencing engine monitoring are as follows:

1. The limits of absolute flight safety may not be exceeded in any way.
2. As no reserve aircraft are kept, the whole fleet must be ready to fly at any time.
3. Within the limits of safety, only economical decisions count as to whether an engine should be changed or repaired on the aircraft, for example, time factor, aircraft availability, cost, etc.

Safety: Minimum Performance—In applying these methods, some thought is required with regard to safety. Along with other factors, which are of no consequence here, safety can be guaranteed:

1. When all precautions are taken to prevent any defect from directly or indirectly endangering passengers
2. When particularly the minimum performance is sufficient for a flight to be concluded without problems
3. When the maximum fuel and oil consumption over the required operational radius is not exceeded

All these safety requirements shall be understood to be the lowest limit of minimum performance (see Figure 1).

Maximum Performance (New Engine)—The layout of the whole propulsion system and the engine design define the maximum possible efficiency. We must be able to expect a maximum life expectancy from the engine parts used. These two factors, together with an optimum module configuration, determine the maximum performance. It is not possible to exceed these limits without changing the engine's basic design. For safety reasons, therefore, an engine must be maintained within the limits of maximum and minimum performance.

ECONOMIC ASPECTS—As economic aspects should determine whether an engine should be repaired while still installed, we must also examine this criterion. An engine repair can be carried out on the aircraft if the overall costs are less than those which

FIGURE 1 Performance envelope.

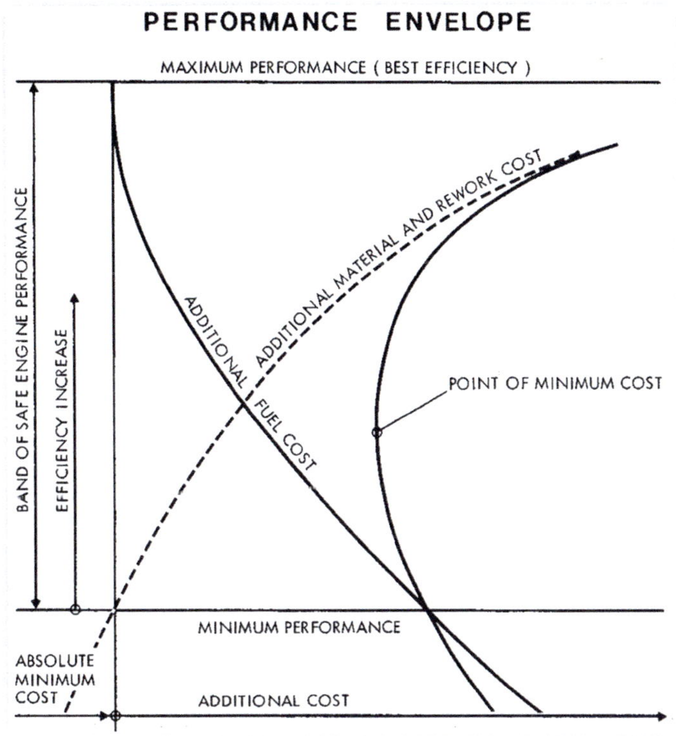

would result if it were to be changed. These overall costs consist of lengthy ground time, large-scale personnel involvement, more complicated inspections, production loss, etc.

The costs of an engine change are based on the actual removal costs, transportation, and often on the somewhat increased use of material for additional preventative action. Most of these financial factors depend upon the structure of the airline and the aircraft involved. Assuming that an engine is not designed completely wrong, the economic operation tends to move also within the limits of minimum and maximum performance. It is our business to find the most economic possible operation.

Economic Optimum—During a repair or overhaul, an improvement can be achieved in the minimum performance by additional means, such as use of new parts, more careful rework, more accurate adjustments, etc. The curve which this performance improvement represents in relation to the additional expenditure can easily be shown in the band between minimum and maximum performance.

A drop in the maximum performance causes an increase in fuel costs, as a result of poorer engine efficiency. In certain cases, secondary damage results from poor quality material, which causes also additional costs. All these expenditures can be shown graphically. The beginning of the curve lies by the maximum performance and shows rising costs with the poorer performance (see Figure 1).

The respective additional costs for each performance condition can be reached by adding these two curves, and as a result of this data, the desired performance can be identified. As fuel and material costs change, the point of optimum economy is also altered with time. Consequently, this optimum point must be formulated separately for each engine.

RELIABILITY CHECKS: AN INSPECTION SYSTEM—Because an economic operation is possible within a relatively wide band and the optimum economic situation is constantly changing, each airline must arrange to check continually the methods of engine monitoring in use. To be unbiased, this examination must be completely independent of the procedure chosen. This can be achieved when only the effects of the monitoring are measured. Typical parameters of this nature are:

1. In-flight shutdown rate
2. Premature removal rate
3. Dispatch reliability
4. Technical delay

The latter two factors influence the image of the airline as regards the passenger, whereas the other parameters should check adherence to the safety limits. By using these parameters for comparison with other airlines, only the effects of monitoring can be measured and not the actual monitoring itself. Within one airline, however, it is possible to compare the various monitoring methods using these parameters.

"Long-Term Engine Monitoring"

The aim of "long-term engine monitoring" should be to operate an engine economically. This will be achieved when each engine finds itself as near as possible to the optimum point on the performance band. All considerations and investigations of this nature are meaningless unless the engine's performance can be sufficiently checked and analyzed. This means that it must be known exactly at which point on the performance band an engine is operating. Ways must therefore be found to analyze the performance of any engine with sufficient accuracy at any time.

MONITORING ON THE AIRCRAFT—This requirement, that is to be able to give performance details at any time, leads to continual and direct measuring on the aircraft. It is unimportant which procedure is used, whether it is the AIDS, the Manual Trend Analysis, or any other system, as long as it is in a position to give the performance condition. The fact that such methods have not been introduced universally shows that it is not easy to fulfill this requirement.

MONITORING IN THE SHOP—One method of engine monitoring, known to everyone but considered less and less, is to be found in the shop. Testing an engine accurately in the test cell after leaving the shop and again before the next shop visit provides detailed performance information. Considering the condition of each single part allows conclusions to be made as to deterioration over a given period. If both lots of information are correctly evaluated for a sufficient number of engines, then for a known starting condition the performance forecast can also be laid down with acceptable accuracy for a longer time period.

"Short-Term Engine Monitoring"

"Short-term monitoring" can be described as the permanent supervision of the functioning capability, by which means damage is able to be detected that could prevent an engine from operating satisfactorily or allow the performance to deteriorate in such a way that it falls below the minimum limit. Included in this category therefore are practically all mechanical problems such as broken blades, failed bearings, and cracked tubes or casings.

Short-term monitoring therefore not only has to keep the engine within the limits of minimum and maximum performance. More important than this, it must also be able to recognize unforeseeable incipient failure.

MONITORING IN FLIGHT—The proper instrumentation of an aircraft must basically make short-term monitoring possible. We now have to correctly analyze the information received from the aircraft or to bring it to where it can be evaluated. In any case it has to be decided whether the next flight can be made or whether the fault must be immediately corrected. Therefore, a lengthy analysis of the information is impossible.

GROUND MONITORING—The normal routine ground inspections complete the "short-term monitoring." The actions taken during these checks are mainly of a preventive nature. Typical examples are:

1. External visual checks
2. Internal visual checks (such as boroscope, X-ray, or gamma ray)
3. System checks (oil analysis and filter checks)
4. Functional checks (ground run-ups)

All these inspections if carried out conscientiously undoubtedly result in a reliable operation. On the other hand, unrestrained application can lead to a tremendous cost increase.

Swissair Monitoring Concepts and Practical Examples

After these explanations, which are mainly general, we should go into the practical procedures of Swissair. It should be explained that fundamental principles used are flexibility in applying the tools available and of the organization involved, together with the most direct information system. It seems expedient to explain the monitoring tools used. The well-known methods such as the various checks need not be specially mentioned. Table 1 gives a general idea of the systems used.

MONITORING: LONG TERM—The intention to automate and improve the manual engine trend analysis system, at the same time introducing the crash recorder according to ARINC 573, led to experimentation with AIDS. Due to the relatively extensive difficulties and cost involved in accurately recording and storing the necessary data, the actual purpose, the development of engine monitoring, was somewhat neglected. The procedure is as follows:

Flight data and engine parameters are taken from the instruments and also from the additional transducers, and are recorded on tape. A printer on board allows the printout of raw data. Exceeded limits and uncorrected takeoff data are immediately and automatically presented.

It is also possible for the printer, when required by the flight engineer, to provide raw engine data for trend analysis. This short-term monitoring tool is, in our opinion, of extreme value.

After the aircraft has landed at the home base, the recorded data is fed into the computer and the raw data then corrected to standard day conditions. Following this, the data can

TABLE 1 Long-term monitoring

Aircraft	Engine type	AIDS	Shop supervision	Manual trend analysis
CV-990	CJ805-23B	-	X	(X)*
DC-8	JT3D	-	-	(X)
DC-9	JT8D	-	X	-
DC-10	CF6-50	X	-	X
B-747	JT9D	X	-	X

* The X within parentheses indicates that these parameters are only used occasionally or in special cases.

CHAPTER 4

be presented as curves or tables (plot and list). A separate program exists and is activated, should given limits be exceeded. A program for engine start and takeoff trend is similarly available. In our opinion, the result of these trends does not live up to the desired expectations.

Although the original purpose of being a short- and long-term monitoring tool has not yet been realized, the AIDS today make problem analysis on the aircraft possible which could not be simulated in the test cell.

Future development should lead to the real AIDS, namely, the Aircraft Integrated Data System. This means that data should not only be recorded but also evaluated on the aircraft. With this, the time-consuming and uneconomic paperwork will be almost totally eliminated.

Shop Supervision—It is vital to possess all details of an engine's history in order to achieve efficient and objective work in the shop. This is accomplished relatively easy by listing problems and faults. To improve the performance condition, however, the efficiency and life expectancy of the main components should be known.

At Swissair, in many cases, the same personnel is responsible for the monitoring as for the engine maintenance, the expenditure for the information system is kept at a minimum. This approved procedure can only be used effectively, of course, when flight experience and shop supervision lie with the same airline. That is why Swissair can no longer apply this method generally.

From Table 1 it can be seen that Swissair uses shop supervision as the only tool for long-term monitoring on the JT8D engines. By laying down stricter test cell limits, it is possible to keep engine performance under control to such an extent that the economic optimum is very rarely not achieved. Comparisons of reliability investigations with our partners show that even using additional long-term monitoring procedures cannot improve the results any more.

MANUAL TREND ANALYSIS: A LONG- AND SHORT-TERM TOOL—As can be seen from Tables 1 and 2, a manual method of engine trend analysis is applied for all aircraft, with the exception of the DC-9. The distinct evaluation as to whether it is used as a long- or short-term monitoring tool for each aircraft type depends on the difference in application of the other monitoring procedures.

By analyzing the curves, the shop possesses data which allows the performance to be determined. This is the definition of long-term engine monitoring, as we see it.

The recording of relatively sudden engine parameter deviations can be considered short-term monitoring and can be checked by the flight engineer under various conditions during flight. At the same time, almost by coincidence, a very reliable implement for instrumentation supervision was found. Should just one single engine parameter show a deviation from the earlier trend, then the flight engineer can try to localize the trouble by operational changes or by altering the instruments. In this way the problem identification by the flight crew is improved in quality because the information has already been proved.

TABLE 2 Short-term monitoring

Aircraft	Engine type	Manual trend analysis	Oil analysis	BSI	X-, gamma ray	Vibration	Run-ups
CV-990	CJ805-23B	(X)*	-	-	-	X	(X)
DC-8	JT3D	(X)	-	-	X	X	(X)
DC-9	JT8D	-	-	-	(X)	-	(X)
DC-10	CF6-50	(X)	X	X	X	X	(X)
B-747	JT9D	(X)	-	X	X	X	(X)

* The X within parentheses indicates that these parameters are only used occasionally or in special cases.

With the introduction of the engine trend analysis, the flight engineer's interest related to engine supervision has been improved, because the trends give him immediate information as to earlier engine behavior and, as a result, a better power of judgment.

ENGINE MONITORING: SHORT TERM—To accelerate troubleshooting and maintenance action, the flow of information is as direct as possible. Problems found by the flight crews are reported directly by them to the line stations and/or home base. Through an HF system introduced in 1972, worldwide direct communication with all stations and all flights is possible at any time. A Maintenance Control Center acts as coordinator between the flight and ground crews.

Allocated to the maintenance are so-called engine troubleshooters, whose work is to analyze the trouble symptoms and find the origin (cause), in this way collecting and evaluating a great deal of information.

When more serious problems arise, the maintenance engineering is always advised and asked for assistance because through daily follow-up of existing problems, consulting the long-term monitoring, and exchanging information with the other partners, maintenance engineering is in a position to survey all important aspects (see Figure 2).

Oil Analysis (SOAP)—For today's jet engines, the earlier methods of oil analysis, where the acid content, viscosity, and contamination were identified, can give little information on the mechanical condition of the oil wetted areas. Therefore, they are no longer in use. Due to the fact that the spectrographic oil analysis can detect exceptionally minimal wear, this procedure can for some problem areas be very successfully used. But SOAP can only be used in cases of slow gradual wear, as the samples cannot easily be taken at shorter intervals than 100 h. Spline wear supervision on the CF6-50 inlet gearbox is a typical example of SOAP being applied successfully (see Figures 3 and 4). Nevertheless, the SOAP is only used on engines which have known problems. However, immediately after a problem has been rectified through modification, the oil analysis is discontinued.

External and Internal Visual Inspections—External visual checks are carried out mainly as routine inspections, within the prescribed requirements. As relatively few important defects are discovered at these checks, we could get the wrong impression that we could quite easily manage without them. However, at Swissair, these checks are a major contribution to the short-term engine monitoring. Additional checks are requested only when specific problems exist.

Frequently experts are brought in from the shop who, because of their specialized know-how, can identify the problems more easily. Thus, for example, an on-wing crack inspection on fan blades is performed by people from the inspection line in the overhaul shop.

FIGURE 2 Information flow.

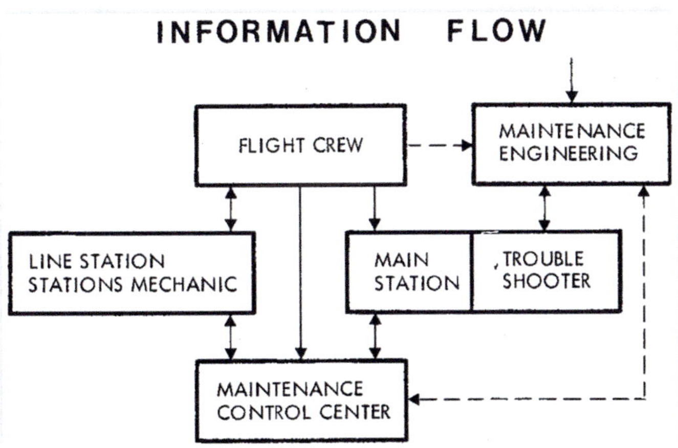

FIGURE 3 SOAP trend.

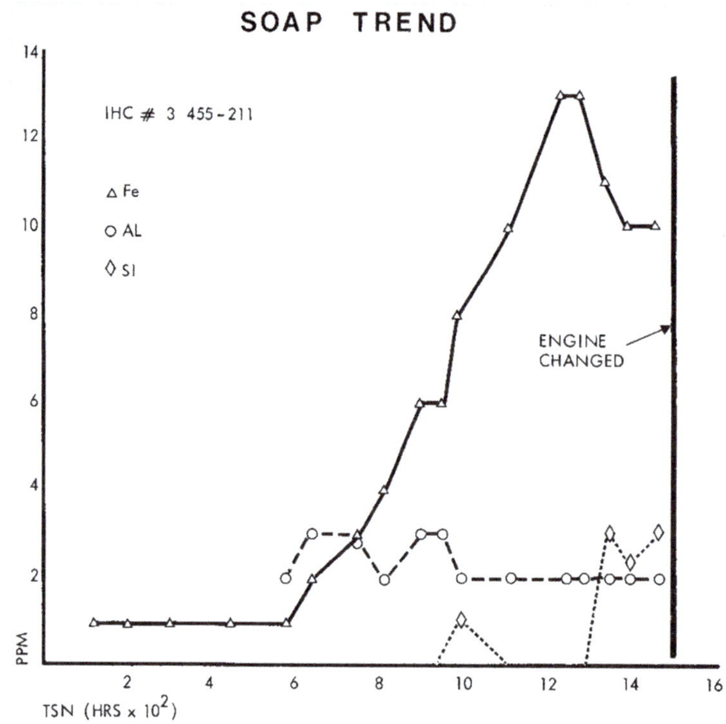

FIGURE 4 Inlet gearbox spline.

INLET GEARBOX SPLINE

The best known method of internal inspection is the boroscope check. It is exceptionally difficult to use this procedure reasonably. To achieve the best results at a minimum cost, we must try to perform boroscope inspections mainly to back up information from other sources.

At somewhat more expense, an X- or gamma ray inspection can be applied on some engines to determine potential problems, but directly because of this increased cost, these inspections are only made in the case of definite complaints or when no other method of inspection can lead to localization of the problem. By using this flexibility for engine monitoring, the expenditure is not only reduced but results are even improved.

Vibration Measurement—During the operation time of an engine, slight changes as regards rotor unbalance inevitably occur. In order to determine detrimental changes, the flight engineer continually enters the indicated values on the engine trend analysis sheet and is thereby always informed on the engine condition. See, for example, Figures 5 and 6

FIGURE 5 Engine trend analysis.

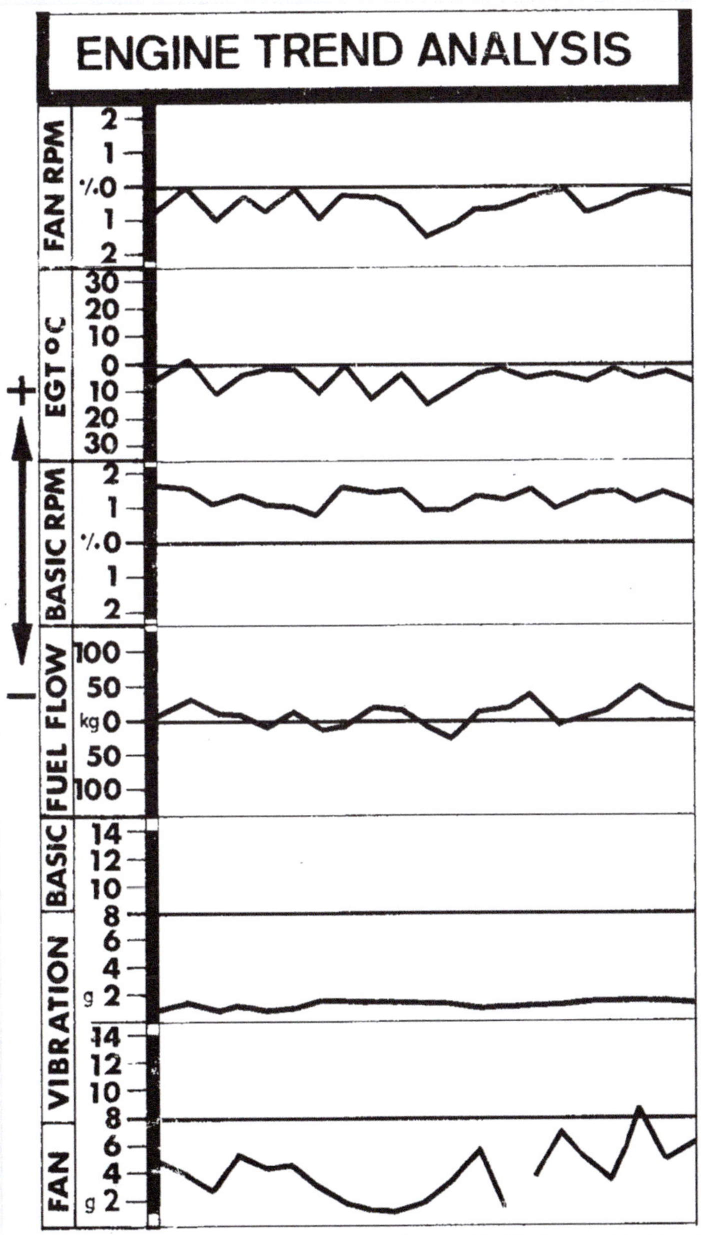

FIGURE 6 Compressor disc.

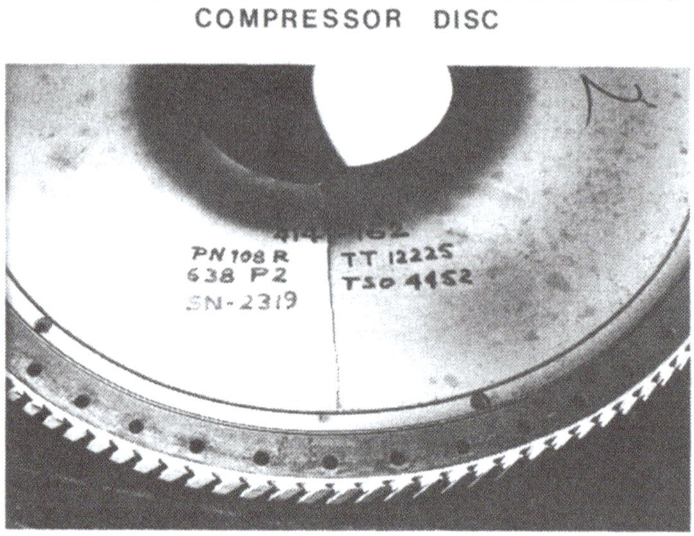

where, thanks to the vibration trend analysis, a cracked compressor disc was discovered in time and as a result more extensive damage and repair costs were prevented. We see an improvement in the quality of the vibration measurement procedures with the use of piezoelectric transducers which are installed as closely as possible to the rotor bearings.

Run-Ups—Almost all engines are pre-trimmed in the test cell prior to installation in the aircraft. A short run-up serves as a trim and throttle alignment check and also as a leak check. After installation, run-ups are only carried out on special request.

Conclusions

These explanations show that a monitoring system as such cannot be considered useful or worthless. The potential of a system is defined by the way it is applied. The application will determine how economical a monitoring system can be. There is a general tendency to overemphasize the value of a monitoring system. Every method of monitoring will be uneconomical, though, if used without restraint or even as an end in itself. Flexibility and controlled selection are the requirements for an economical monitoring system and in this lies the main difficulty of the new automated and computerized methods.

In our opinion, an engine monitoring procedure must embrace all areas such as operation, maintenance, overhaul, and test cell, if it is to be applied effectively. Technical equipment is a necessity, but the flexible use of the specialists' know-how invariably determines the success or failure of engine health monitoring.

The so-called "Short-Term Engine Monitoring" on the basis of today's problems with jet engines has definite priority over the long-term monitoring. Whether economical automation of this short-term monitoring is possible with today's technology appears rather questionable.

The aim of future development of the so-called "Long-Term Engine Monitoring" must, due to the application of the new Engine Modular Maintenance Concept, be able to provide more information on each individual module rather than on the engine as a whole. It is obvious that, due to the different specific circumstances, each airline has to find its own way of engine health monitoring.

CHAPTER 5

Modular Engine Maintenance Concept Considerations for Aircraft Turbine Engines

David B. Edmunds
Propulsion Performance/Stability Div., Wright-Patterson Air Force Base

Maintenance on gas turbine engines results from failures in one or more of the engine's major subsections. Maintenance and support costs can be reduced if the capability exists to isolate the failures down to the lowest possible component in the engine. The modular engine maintenance concept has recently been evolved to facilitate this cost reduction. The important considerations as well as the benefits of this concept are discussed. Considerations included are (1) design, (2) failure modes and isolation techniques, and (3) implementation. The Air Force F100-PW-100 engine maintenance concept is used for discussion purposes.

BECAUSE OF INCREASING ACQUISITION and support costs of advanced turbine engines, the Air Force is investigating new maintenance techniques that contribute to reducing the total cost of ownership. Some of the more significant of these techniques are (1) increased on-condition maintenance, (2) inflight engine monitoring systems, and (3) modular engine maintenance.

A turbofan engine may be viewed as consisting of several basic units: a control, accessories, and the turbomachinery or gas path elements. The turbomachinery can be divided into individual modules consisting of the fan, compressor, combustor, high-pressure turbine, and low-pressure turbine. A multitude of failure modes can occur in use which will result in the entire engine being removed for repair. The benefit of the modular maintenance concept is derived from the ability to repair only the module in which the failure has occurred. This paper will discuss considerations involved in the implementation of the modular engine maintenance concept.

Design Considerations

The modular engine maintenance concept can be viewed from three aspects: (l) engine design philosophy, (2) failure modes and isolation techniques, and (3) implementation. The design philosophy aspect considers the ease with which a module can be separated from the other modules for maintenance purposes. The capability to disassemble a turbine engine by major subsection has existed on most engines even though they were not designed as modular engines. For example, the Air Force J75 engine produced before 1960 without modularity as a design consideration is under certain circumstances maintained as if it were a modular engine. The high-pressure turbine can be removed in the field, shipped to the depot for repair, and replaced by another high turbine section. The same can be done with the other major subsections except in cases where balancing is required.

The Air Force F100-PW-100 engine, first produced in 1974, was designed to be maintained as a modular engine. The F100 engine consists of five modules as shown in Figure 1: fan, core (the high-pressure turbine can be separated), fan drive turbine, augmentor and main, and gearbox. As with the J75 any of these modules can be removed, replaced, and shipped separately. Beyond this similarity there are some important differences in the design and maintenance concept of the two engines. The high-pressure turbine on the F100 can be removed in about 1/3 the time of the J75. This time savings is the result of designing the plumbing lines to be readily accessible for disconnect at the module split, a simple unbolting of the case between each module, minimum balancing requirements, and the inclusion of the bearings as part of the individual modules.

Another difference is the design of borescope ports, and temperature and pressure instrumentation ports at strategic locations on the F100 which allow the maintenance personnel to isolate a problem to the module level. Without properly located instrumentation the gas path analysis techniques discussed in the next section could not isolate performance problems to the individual module.

Whereas the J75 and F100 engines can both have modules swapped in and out, the J75 is logistically managed and tracked as a whole engine with a fixed maximum operating time (MOT) between overhauls of the engine. The F100 is not tracked as an engine but as five individual separate serialized modules. The engine serial number is used to locate where the various modules are and to credit accumulated time and cycles. Life-limited parts in each F100 module were identified during development testing and through various structural integrity tests. These limits are used to determine when a part should be removed from the module. The modules remain in use (but not necessarily in the same engine) until the accumulated usage exceeds a limit or some other condition necessitating removal occurs (e.g., burned turbine nozzle, low performance).

FIGURE 1 F100 engine modules.

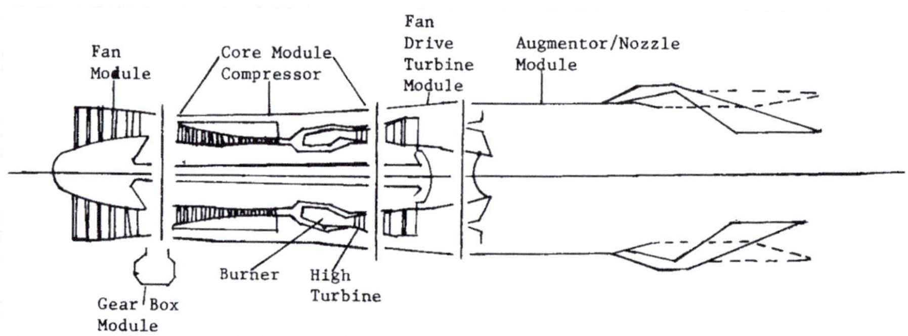

Failure Modes and Isolation Techniques

The failure modes encountered by the gas path modules can be divided into two general categories: (1) those that affect a change in the performance and (2) those that affect a change in the structural integrity of the module. These two categories certainly overlap as shown in Figure 2. A failure mode such as excessive tip clearance affects the performance of the module but not necessarily the structural integrity. Fatigue cracking of a blade or rotor disk affects only the structural integrity of the part, but a burned or warped turbine stator can affect both the performance and structural integrity. A failure is defined in this paper as the inability of a part to perform its designed function within specified operating limits.

Present-day turbine engine maintenance practices make use of extensive measurements obtained both on the ground and inflight. The basic purpose of these measurements is threefold: to predict, detect, and diagnose the two categories of failure modes presented above. These measurements are used in a variety of ways to perform these purposes and can be classified as direct or indirect.

The indirect methods of measurement are typically used to determine the structural integrity of a module. In the past, engine operating time was tracked and when a predetermined MOT was reached the entire engine was overhauled. This approach is considered indirect because it assumes that all engines were operated to a certain set of conditions and that material properties were the same. More recently, rotor transient cycles and time at temperature have been tracked in an attempt to better account for the variability in usage between engines due to different missions and pilot techniques.

The structural integrity or life remaining for a particular engine or module cannot be obtained through a direct measurement. Visual inspections can certainly determine that parts have large cracks, but if a crack is present a failure to some degree has already occurred. For example, fatigue is an accumulative failure mode for which only the last stages of failure can be detected through visual means.

The failure modes affecting performance can generally be determined by direct methods. In the past, the most common method of determining these failure modes made use of inflight out-of-norm functional occurrences. Typical occurrences were compressor stalls, or predetermined operational maximum limits on one or two overall

FIGURE 2 Failure modes.

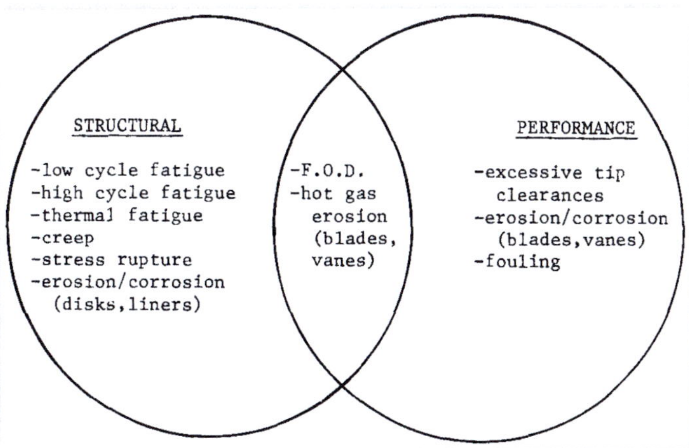

engine parameters such as exhaust gas temperature (EGT) or fuel flow. A variation on this approach was to trend overall engine measurements as a function of time. A shift in these measurements from their nominal operating level was used as an indicator of some sort of performance degradation in the engine.

Although the above methods make use of measurements to directly determine failure modes affecting performance, they do not isolate or quantify the failure to a particular module but generally apply to the entire engine. The capability to isolate a failure to a specific module forms the basis of the modular engine maintenance concept. Whereas the capability to account for structural failures primarily depends on indirect methods, recent developments in gas path analysis [1, 2, 3] have provided the capability to isolate and quantify failures affecting performance to the module level implicitly. To be implicitly detectable (i.e., implied from their effects on the measurable parameters) the failure modes must clearly be of a nature and magnitude that will produce an observable change [1].

The performance of the individual engine gas path modules can be represented by certain basic performance parameters. Failure modes affecting performance and associated with the fan or compressor will manifest themselves as changes in either the air pumping capacity or the adiabatic compression efficiency or both. Likewise failures in the turbine modules will manifest themselves as changes in either the turbine effective nozzle area size or the adiabatic expansion efficiency or both. These independent parameters are not directly measurable, but using dependent measurements such as temperatures, pressures, fuel flow, and rotor speeds throughout the engine, quantitative changes to the independent parameters can be implicitly determined through the use of gas path analysis techniques.

Implementation

The preceding sections of this paper discussed the design, failure modes, and analysis considerations necessary for implementation of the modular concept within the Air Force, and specifically to the F100 engine. The two basic maintenance concepts used on the F100 engine are the on-condition concept and the modular concept. On-condition maintenance is a concept that dictates maintenance action based on the actual condition of the installed component, module, or engine as determined by visual inspection, measurement, test, or other means without a teardown inspection.

Turbine engines to some extent have always been maintained on-condition. The important improvement on the F100 is the elimination of the fixed time engine overhaul schedule. The modular concept in conjunction with the on-condition maintenance concept provides the framework under which only the parts experiencing a failure are repaired.

Obviously if all the modules in an engine failed at the same rate, there would be no benefit to the modular engine concept. In actual practice there is a considerable disparity in the failure rates of modules. By supplying to the field the proper mix of modules based on their relative failure rates, the maximum number of engines can be kept in operation with the minimum complete spare engines. Benefits will result from mixing the buy of spare modules as a function of their respective removal rates. Also the transportation costs will be less when only bad modules are shipped as opposed to shipping a whole engine with one bad module and four good ones. At engine maturity, F100 modules returned to the overhaul facility in whole engine configuration are expected to account for only 5% to 10% of the modules returned.

To obtain the above benefits requires extensive base-level maintenance facilities. These facilities, referred to as Jet Engine Intermediate Maintenance (JEIM) shops, must have the equipment to remove and replace all the modules. The base level must also have the necessary tools to isolate faulty modules whether for performance or structural failures. This includes engine ground test stands, installed trim facilities, borescope, ground diagnostic equipment, and a gas path analysis program. These base-level facilities will typically service 175 F100 engines. The majority of repair work on the modules will be at a single overhaul facility.

As previously discussed many of the failures that occur in turbine engines are isolated through the tracking of life-limited parts, on-condition occurrences, and the use of ground testing and inspection equipment. The capability to isolate performance problems to the module level will also have to exist at the base level for the F100 engine. At the current time this capability does not exist in the Air Force and the following discussion will detail its anticipated implementation.

Two basic assumptions are made in describing the implementation of a module performance determination capability. First, a gas path analysis program must be available with the capability to quantify the performance parameters of the various modules. It must be small in computer memory requirements so it can reside at the base-level engine shop. The output must be in a form that can be interpreted by base maintenance technicians, as performance engineering personnel are not available. A second assumption is that sensors can be installed in the engine at the proper locations to provide the gas path analysis program the required accurate measurements. (These locations are shown in Figure 3.)

Initially the gas path measurements will be obtained from ground operation either on a test stand or installed in the aircraft. In the future the measurements may be obtained inflight (in order to reduce ground operating time and fuel required) with the aid of an engine diagnostic system (EDS) currently being developed for flight evaluation. Careful attention must be given to gathering data inflight. Among the factors that require consideration are (1) engine stabilization times, (2) installation effects, (3) nozzle position, and (4) Reynolds number effects. These factors can introduce large errors in the analysis technique or provide invalid conclusions if not handled properly.

Ground run-up measurements will be obtained whenever the engine control needs to be trimmed. Engine trims are performed on-condition. Typical conditions requiring trim are stalls, augmentor blowouts or no-lites, pilot squawked low performance, control malfunctions, and major gas path component or control replacement or repair. If the engine control cannot be adjusted to provide the specified maximum non-afterburning performance with the engine operating below a maximum allowable turbine temperature limit, the engine will be rejected. The cause of the problem must then be isolated to either a control malfunction, sensor failure, or gas path deterioration of one or more modules.

FIGURE 3 F100 gas path sensor locations.

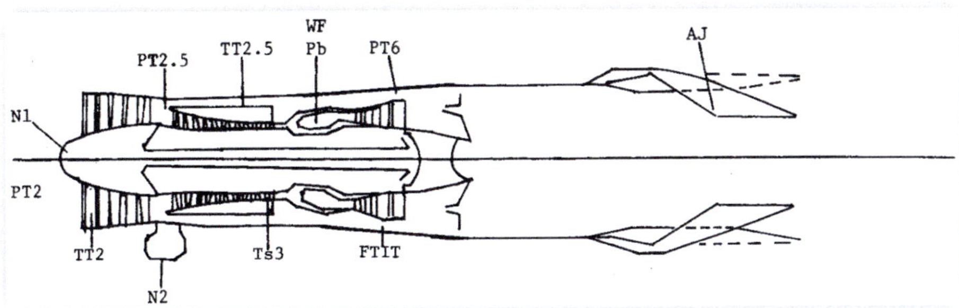

The measurements obtained from the trim run will be input to the base-level gas path analysis program which will have the capability to determine the performance levels of the individual modules relative to the nominal new engine baseline. The program will convert the individual module performance parameters into its relative contribution to fan turbine inlet temperature (FTIT) increase above nominal. For example, if an engine requires 50°F more turbine temperature than nominal to make the specified performance, as determined by engine pressure ratio (EPR), the high-pressure turbine might account for 40°F of this and the high-pressure compressor 10°F. For this example the apparent decision could be to replace the high-pressure turbine. The actual decision-making process will consider the gas path analysis information in conjunction with the time remaining information on the life-limited parts and a visual (borescope) inspection of the engine. This decision-making process would take place, for most cases, at the base level by base personnel. For special cases the overhaul facility will be consulted for additional information. The typical maintenance process and information flow is shown in Figure 4.

FIGURE 4 Typical maintenance process and information flow.

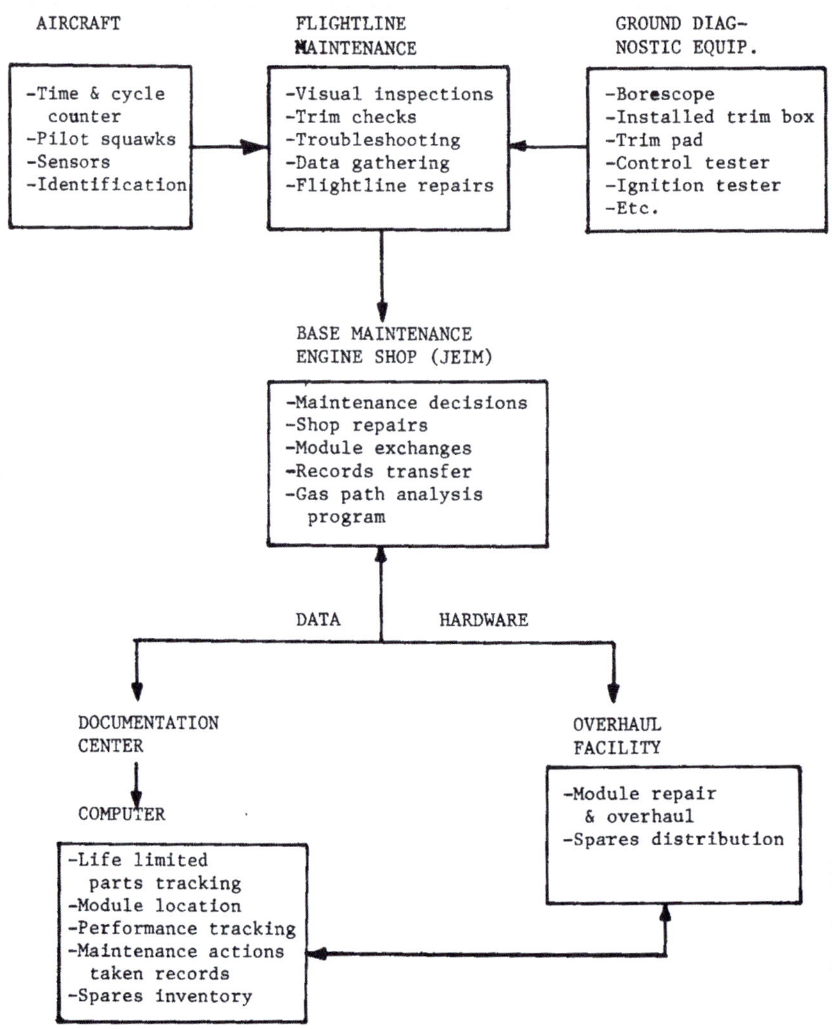

Another consideration is the determination of performance levels for modules in the inventory that are either new, recently repaired, or previously installed in another engine. Modules returned to the overhaul facility will be repaired as required and returned to the inventory. Dimensional and functional checks would be performed during reassembly of a module; however, there presently is no operational or performance test required for new or repaired modules to assure proper operation when installed with other modules. This means that when a module is installed in an engine, the performance level or capability of that module is unknown. One solution to this problem, commonly referred to as "mixing and matching" of modules, is to assign a performance level number to each module in the inventory. This performance number could be based on its contribution to the engine's FTIT increase relative to a common performance baseline. New modules could be assumed to have 0° FTIT increase. Repaired modules could have FTIT margins assigned dependent on the extent of the repair and its past performance level. The performance level for used modules could be obtained from their last trim run as described previously. These performance numbers could be tracked for each module the same as cycles are presently tracked and used to decide which module to use when building up an engine. A basic assumption of this approach is that a specific module performance will be the same when installed in different engines. Another assumption is that the gas path analysis program is capable of calculating performance levels of the modules and converting these levels to temperature margins with a satisfactory confidence level.

Summary

The modular engine maintenance concept offers great potential for cost of ownership savings in reduced overhaul costs, spare engine requirements, and transportation costs. The potential savings are enhanced when the initial engine design incorporates the modular concept. To obtain these benefits requires a significant change in the Air Force methods of maintaining turbine engines. The record-keeping requirements are significantly increased. The capability must exist to isolate failure modes to a particular module of the engine. This capability should exist at the base level and include analytical and inspection techniques. The failure modes can be classified as either affecting the structural integrity or performance output of the modules. The structural failure modes are usually determined through indirect means and the failures affecting performance through direct methods such as gas path analysis. The gas path analysis techniques must be capable of determining performance levels of individual modules. These calculated performance levels will be used in deciding which modules require repair and aid in the module buildup of engines.

References

1. Urban, L.A., "Gas Path Analysis Applied to Turbine Engine Condition Monitoring," AIAA Paper 72-1082, December 1972; *Journal of Aircraft*, 1st ed., 10, no. 7 (July 1973): 400-406.
2. Urban, L.A., "Parameter Selection for Multiple Fault Diagnostics of Gas Turbine Engines," ASME Paper No. 74-GT-62, April 1974.
3. Kos, J.M., "Multiple Fault Gas Path Analysis Applied to a Twin Spool, Mixed Flow, Variable Geometry, Turbofan Engine," Report No. HSER 6794, Hamilton Standard Division, United Technologies Corporation, October 1975.

CHAPTER 6

Developments in Performance Monitoring and Diagnostics in Aircraft Turbine Engines

Carolyn Smith and Ronald DeHoff
Systems Control Technology, Inc

Martin Huffman
Air Force Wright Aeronautical Laboratories/POTC

Automated performance monitoring of aircraft turbine engines can be characterized as a three-phase process which includes data acquisition, data processing, and maintenance information management. With recent research and development activities in these areas, more sophisticated methods for performing these tasks have been developed. Results show promise for improving the reliability, efficiency, and utility of automated performance monitoring and diagnostic systems.

IN ORDER TO ACHIEVE a reliability-centered maintenance (RCM) policy for aircraft engines in the Air Force, methods and systems have been developed which monitor the health of modern propulsion systems. Information derived from inflight collected data is analyzed and reduced to meaningful engine health indicators at the level of the engine module. This information, in concert with other engine data and maintenance information, must be integrated and made available to maintenance decision makers in a fashion which supports RCM. The performance monitoring/diagnostic process can be characterized as a three-step process which includes data acquisition, data processing, and maintenance information management. Each step is unique and critical to the success of the overall maintenance policy.

CHAPTER 6

Data Acquisition

The data acquisition portion of the engine performance monitoring/diagnostic process includes the measurement and selective recording of engine data during flight. This data must support the detection of inflight malfunction events, the tracking of engine life usage, and the assessment of long-term performance degradation.

The Advanced Propulsion Monitoring System (APMS) Program is notable in the way in which it addresses data acquisition problems. Under the support of Air Force Wright Aeronautical Laboratories/POTC, Systems Control Technology, Inc., and General Electric Company (AEG) are building a prototype inflight engine monitoring system. The program will demonstrate a prototype configuration representative of a flight-type installation. System test will be conducted on a GE F101-DFE turbofan engine in an altitude cell. The test system configuration is illustrated in Figure 1.

The system is unique in the way in which it collects data for trending purposes. Many of the current performance monitoring systems rely on manual or ground-based automatic trending and trend plot methods. In the APMS, initial steps of performance trend monitoring can be executed on board simultaneously with data acquisition functions. A trending algorithm has been designed which automatically records engine data in a "Trend File" when the aircraft has reached a stable flight point. The logic first checks to see that data falls within the cruise data acquisition window, which is based upon power level angle, ambient pressure, and Mach number. Engine stability is determined according to whether significant deviations are present in the sequential short-term and long-term averages of engine parameters. When deviations are within preprogrammed limits specified by the engine manufacturer, data is saved. If an instability or engine

FIGURE 1

APMS Test System Configuration

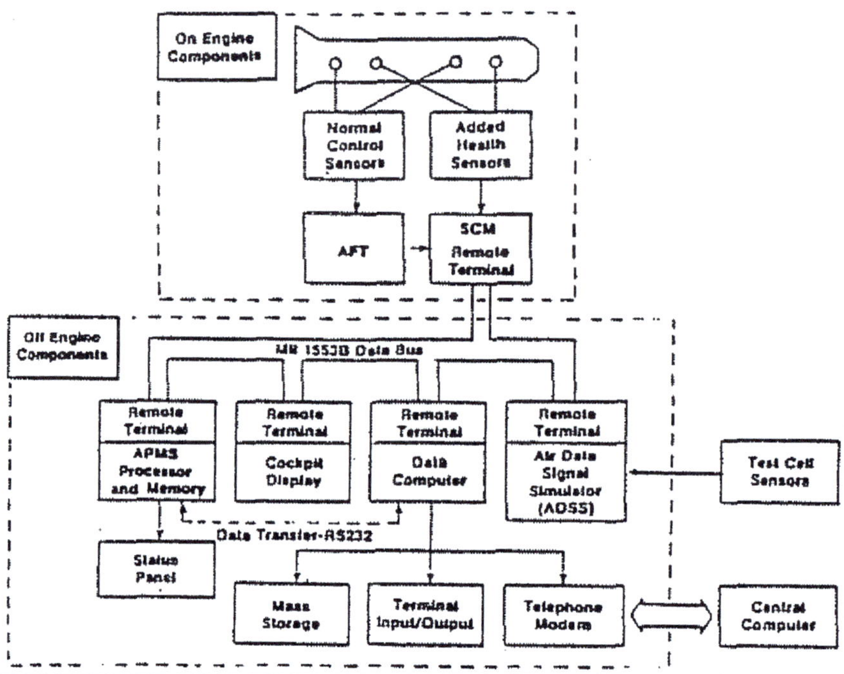

exceedance event is detected, data saves are terminated. This acquisition logic saves time by eliminating manual recording and plotting, and saves space by performing initial trend filtering functions prior to recording.

Experience in inflight engine data analysis [1, 2] has shown that data acquired during aircraft takeoff is statistically more repeatable than data collected during other portions of the flight envelope. This characteristic makes takeoff data amenable to use in a trend analysis procedure which relies on thermodynamic engine models. Data acquisition software logic has been designed which provides for automatic recording of takeoff data and subsequent storage in a "Takeoff File." Increased reliability of performance trend monitoring should result.

Engine usage monitoring is enhanced by the use of a Mission Profile Algorithm. Mission definition information is recorded by this logic for purposes of off-board analysis of cyclical and stress-related degradation phenomena. Corner point data sets are saved when checks of fan speed and fan inlet temperature indicate out-of-limit or transient behavior. This is accomplished by a series of software modules which check individual readings against the median of the previous five readings, determine the mode to be "stable" or "unstable," and distinguish between fast and slow rates of excursion. Figure 2 is a plot which illustrates this economical method of mission reconstruction.

The quality of data collected by the data acquisition system is critical to the effectiveness of the performance monitoring/diagnostic procedure. Sensor error modes including bias, noise, drift, and hard failure must be considered in the design of the data acquisition logic. Key data quality issues center upon around the extent to which data should be filtered prior to recording and the level of sophistication with which failed channel readings should be accommodated.

Sensor diagnostics routines have been designed that detect, isolate, and accommodate failed channel readings. The logic structure is table driven and static linear models are employed in both detection and reconstruction. These features increase the execution speed of the software and economize storage and space requirements. The software has the ability to account for sensor noise, engine-to-engine variation, engine

FIGURE 2

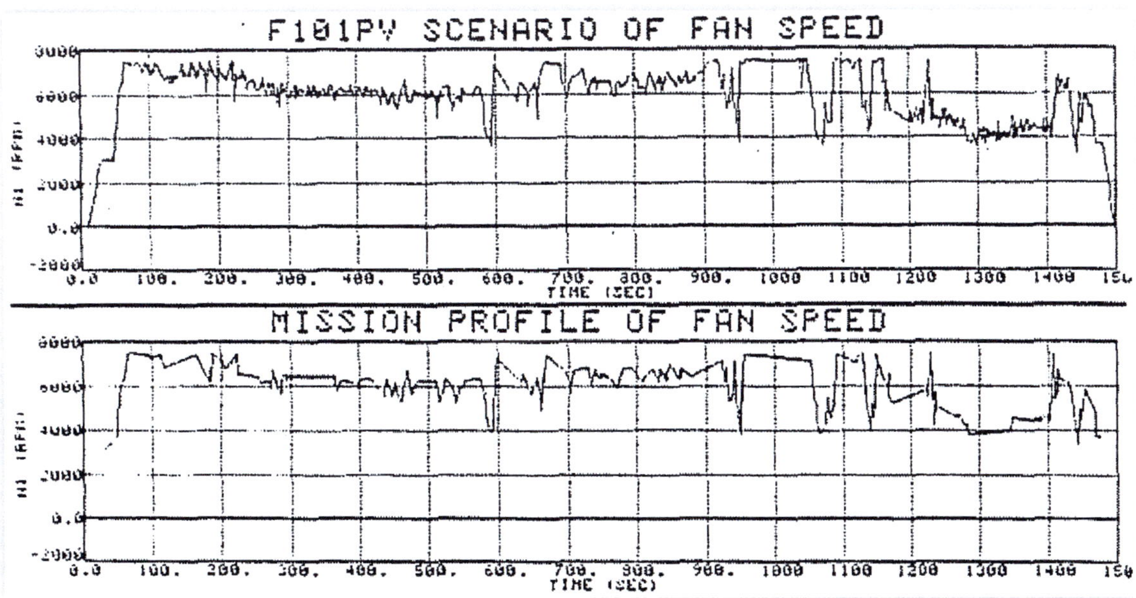

transients, and need for complete model replacement. To accommodate sensor noise, a statistical threshold exceedance test is employed to flag inconsistent readings. Model parameters and tabular input to the isolation and reconstruction logic are part of a "load deck" which is a replaceable part of the APMS software. Thus, a flexible framework is provided which can be tailored to specific engines and sensor sets.

Data Processing

While data acquisition involves the measurement and recording of physical parameters in a harsh dynamic environment over an extended period of time, data processing must account for errors in the measurements and reduce the data to operationally significant quantities. Under support of the Air Force Wright Aeronautical Laboratories, Systems Control Technology, Inc. and the McDonnell Aircraft Company have focused on these issues in the Turbine Engine Fault Detection and Isolation Program. Using data acquired by three inflight monitoring systems, data processing techniques were developed which derive estimates of engine performance from measured data. The monitoring system data used included data collected by the TF34-TEMS, the F100-EDS, and the TF41-IECMS.

A variety of techniques are used to reduce measurement errors. Random error due to sensor and channel noise can be reduced by averaging. Algorithms which filter data and use groups of data points taken at different parts of the operating envelope reduce errors further. Analysis of signal noise establishes the optimum number of sampled points. Engine models can be used to validate a set of measurements to detect and isolate sensor failures.

In addition to sensor accuracy problems, engine build variations and hardware changes (module swaps) during the engine lifetime must be addressed. Calculation of the operating baseline and sensor calibrations can be performed after each maintenance action. Alternately, a reinitialization of the confidence levels in the filtering algorithm can discard information about unaffected parameters.

To be a useful part of the maintenance cycle, the processed performance monitoring data must reflect engine health at the engine module level. Deteriorated or failed modules are isolated at the intermediate (base) support level and maintenance and inspection procedures are geared to identify work items required to restore the component to operational condition, or specify shipment to the depot level maintenance area. Generally, deviations in measured parameters are due to shifts in engine behavior caused by deterioration and failure. This can be quantified by changes in the areas and efficiencies of engine components known as fault parameters. Measured deviations are compared against fault parameters, and inferences concerning engine health are drawn. This sequence transforms measurements to health assessment indicators.

Estimation techniques have been used to assess the effect of small deterioration processes on overall engine performance. A thermodynamic cycle monitoring algorithm was developed to accomplish this end. Trending data is analyzed by a sensor diagnostics routine which uses engine models to isolate hard and intermediate sensor failures. Synthesized data is generated for failed channels and received by a health index estimation routine, which estimates engine health status. Current engine health index estimates are input to a trending routine to be analyzed with respect to previous estimates for statistical changes.

Figure 3 illustrates the results of the trending algorithm. It shows a plot of a net engine health rating and a core rating as a function of total engine operating time. The sharp increases in the trend plot at 165 and 250 h can be correlated with engine maintenance. Similar results were seen in other engine data sets. The plot confirms that an

FIGURE 3 Trending algorithm results.

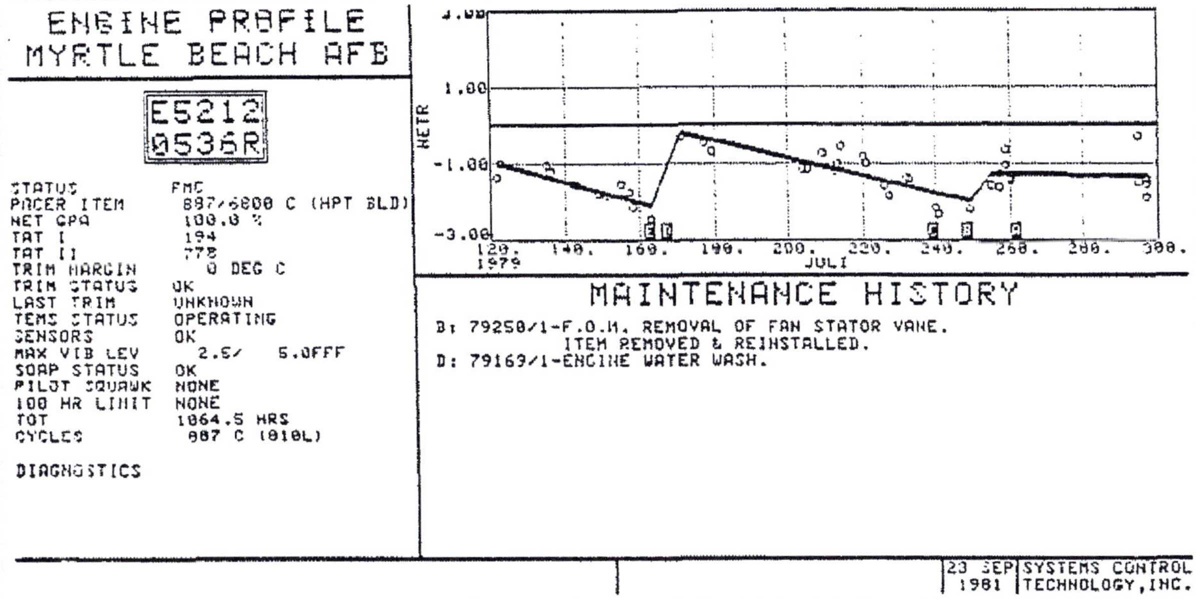

algorithm can be developed which reduces engine monitoring data to operationally significant module health indices. This plot suggests the use of the analytical method for the purpose of predicting or scheduling maintenance actions. Conceivably, lines drawn to fit trend plots could be extrapolated to predetermined rating levels, at which point in time a maintenance action could be scheduled.

Data Display and Maintenance Information Management

The third aspect of automated performance monitoring involves output display and the integration of data from various sources with inflight acquired data. The data must be presented in a meaningful format such that an effective decision support system is offered to the maintenance personnel. This system must promote the use of RCM policies.

Under the Turbine Engine Fault Detection and Isolation Program, a Maintenance Information Management System (MIMS) was developed. Data products can be generated which have been designed in response to requirements of the engine management community. MIMS generated data products include:

1. Daily base status summary and exception reports
2. Engine profile and diagnostic information
3. Detailed plots and tabulated displays for oil analysis data, vibration data, maintenance history data, life usage, and performance trending data for specific engines

The base status summary is intended for daily use by base level management. It updates aircraft and engine status and documents critical changes in engine condition (Figure 4). The summary shows the breakdown of aircraft by mission capability and

FIGURE 4

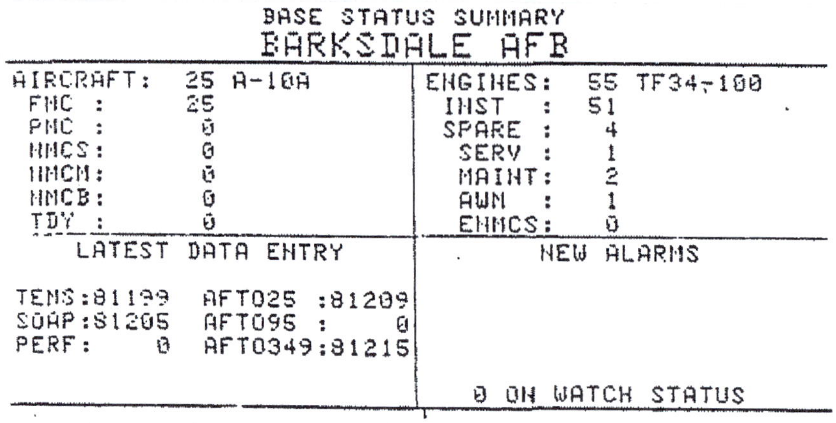

identifies aircraft without engines. For uninstalled engines, it provides the distribution of engines in the local repair pipeline. An estimate of the backlog of maintenance man-hours is provided for engines awaiting maintenance (AWM) and in maintenance (MAINT). A critical change of engine condition is indicated using an alarm. Watch status indication is provided for significant, noncritical shifts or trends in monitored performance factors. The engine profile display (Figure 5) contains summary entries, status/documentary information, and diagnostic messages. It provides a snapshot of summary information for an individual engine.

Data items which can be accessed with this display product include life usage factors (time or cycles), overall performance, trim margin, and current vibration levels. Documentary information includes base location, engine serial number, location, and date of most recent trim.

FIGURE 5

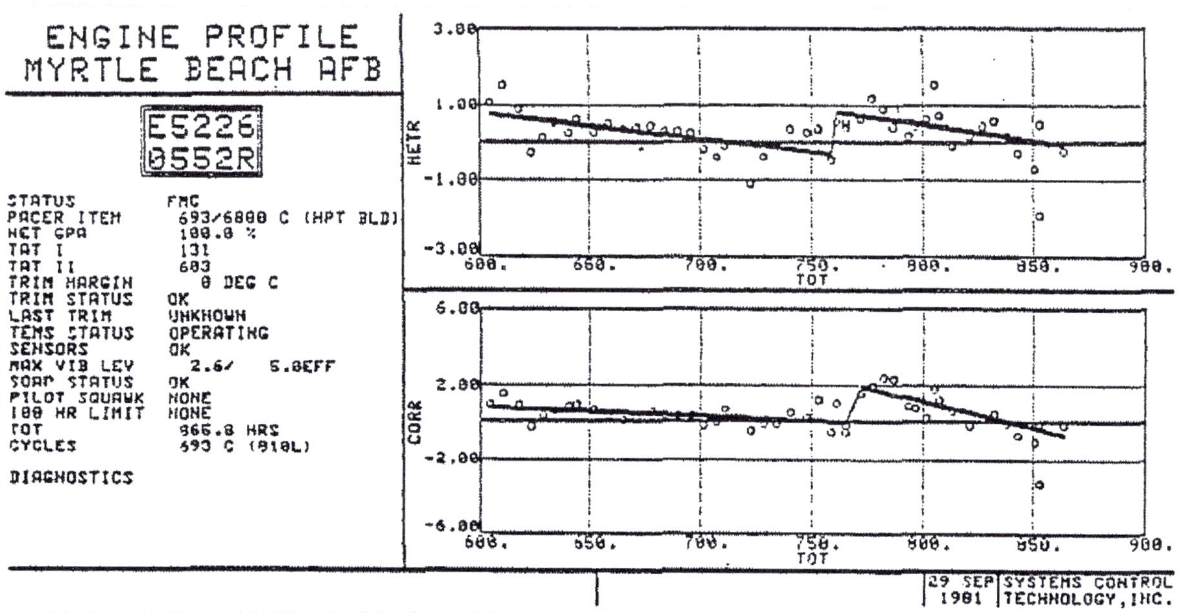

A second-generation MIMS is being designed which is to integrate engine diagnostics and trending data with other engine data. The Comprehensive Engine Management System (CEMS) Increment IV is an outgrowth of the Air Force's implementation of the RCM Concept for propulsion engines.

The overall CEMS (D042) is being implemented by the Air Force Logistics Command in four increments, each of which addresses a maintenance management functional area. Increments I and II provide the TCTO portion of Configuration Management for the TF34, TF41, F100, TF33, TF30, J85, J60, and TF39 engines. Increment III provides status reporting, pipeline analysis, financial accounting distribution, inventory control, and actuarial analysis and requirements computation for all engines integrating existing Air Force data systems into a single engine management system.

Under support of the Air Force Logistics Command, Systems Control Technology, Inc. is developing the CEMS Increment I? (CEMS IV) program. CEMS IV at the base level provides engine diagnostic and trending capability to the entire Air Force inventory of jet engines. CEMS IV at the Central Data Bank (CDB) level provides centralized collection and analysis of engine performance, condition, and diagnostic data.

The CEMS IV base system provides engine maintenance personnel with performance, diagnostic, and trending information to support the application of On-Condition Maintenance (OCM). The diagnostic and trending information is generated by analysis routines from engine performance and inspection data, including oil analysis and significant maintenance action data. The analyses produce health indices, diagnostics, performance trending, and historical maintenance information allowing the engine manager or propulsion branch chief to quickly determine the level of performance of an engine. Specific requirements include:

1. Maintain parameter values for each engine type
2. Gather, store, and process nondestructive inspection data
3. Gather and maintain engine performance profiles
4. Plot specific performance parameters of given engines over a period of time
5. Establish engine health data containing profiles of individual engines
6. Provide analytic routines to develop output products from the engine health data base
7. Provide presentation products for performance data, health profiles, ranking, and significant maintenance action data
8. Provide performance data, health profiles, ranking, and significant maintenance action data to depot personnel

The program completely supports the SAC Engine Condition Monitoring Program (ECMP), the base Joint Oil Analysis Program (JOAP), and the A10 Turbine Engine Monitoring System (TEMS).

The base system is an interactive, real-time information system, providing users with online query and analysis capabilities to support base engine maintenance. It passes data to the CEMS CDB. It is structured in three levels. Level A, Level B, and Level C are nested so that each higher level includes all the capabilities of lower levels. Level A provides the ability to monitor, analyze, and trend oil analysis data (JOAP). Level B includes the capabilities of Level A and provides the ability to monitor, analyze, and trend performance data collected manually by the flight crew in the SAC ECMP environment. Level C provides performance analysis of data collected by automated inflight monitoring systems such as the A10/TEHS or F101/APMS.

The CDB portion of CEMS provides major commands and headquarter operations with performance-derived management information to support application of OCM, early identification of material deficiencies, and engine management requirements. The management information is generated by analysis routines from basic performance data, gathered by base level monitoring, and accumulated in the CDB. The performance analyses produce health indices, diagnostics, and performance trending information, which allows the command level engine manager quickly to determine the level of performance, readiness status, and the quality of maintenance actions. This determination is necessary to support the Air Force's worldwide implementation of OCM programs for propulsion engines.

The functions of the CDB are similar to those outlined for the base level CEMS except the CDB is to process data on a global, force-wide level for integration into a central information and management system. Figure 6 shows the basic architecture of the CDB system and its interface with the main software modules. Interaction with the data base and the data communication terminals is through IMS data base software on an Amdahl V7B triplex at Tinker Air Force Base, Oklahoma.

Sample CEMS displays include a report generator, statistical profiles, ranking displays and trending, extrapolation, and linear plots. Figure 7 is a list of reports which can be generated by CEMS. By entering a report number, the system user can generate a table, plot, or list as specified by the report definition. Figure 8 shows a statistical summary of oil iron content by engine in a typical tactical fighter group. Figure 9 ranks parameters in decreasing order, while Figure 10 summarizes the statistics for groups of engines. Historical trending, extrapolation, and linear plotting capabilities are illustrated in Figure 11.

FIGURE 6 CEMS IV prototype architecture.

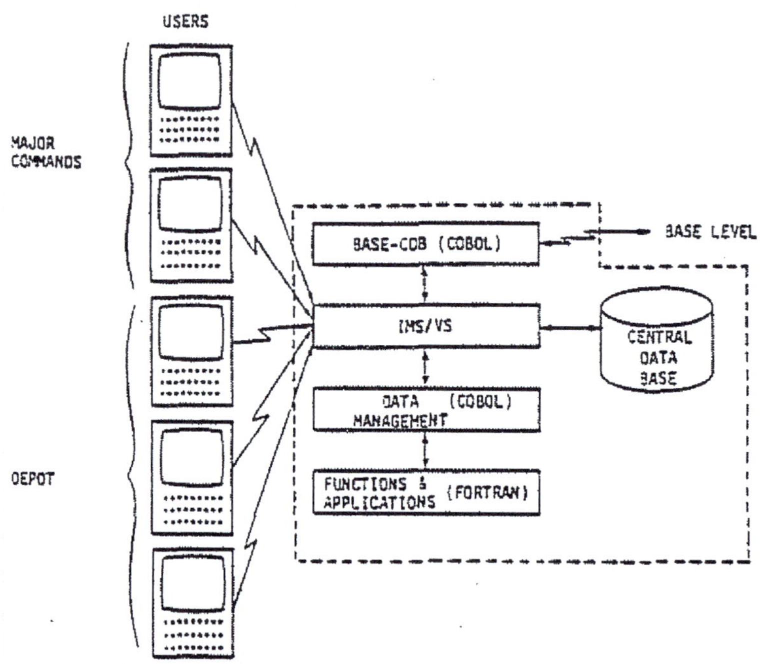

FIGURE 7

	REPORT	EXECUTES COMMAND	AND CAUSES
1	R01	TAB,MODE,NFRL,NFTR,IDLE,VGSC,NRECS=50,ESN=EACH	
2	R02	TAB,TSO,TOIL,COIL,FE,AG,AL,CR,CU,NI,COB,S,ESN=EACH	
3	R03	PLOT,NG,NF,EOT,MODE=2,2,ESN=EACH	
4	R04	SET,ESN=5859	R05
5	R05	PLOT,FE,JULI	R06
6	R06	PLOT,NI,JULI	R07
7	R07	PLOT,POIL,JULI,MODE=2	R08
8	R08	PLOT,POIL,JULI,MODE=4	
9	R09	MPL,CBC,FFC,EFC,CBF,FFF,EFF,RECN,MODE=2,NRECS=30,6,N,ESN=EACH	
10	R10	STATS,POIL	
11	R11	RANK,TSO	
12	R12	RANK,NIPR.THD,A	
13	R13	GROUP,TSO,MINB=200,MAXB=2000,NBINS=6	

DISPLAY REPORTS

REPORT DEFINITIONS		
917TH TFG	TF34-GE-100	02JUL15(1606)

FIGURE 8

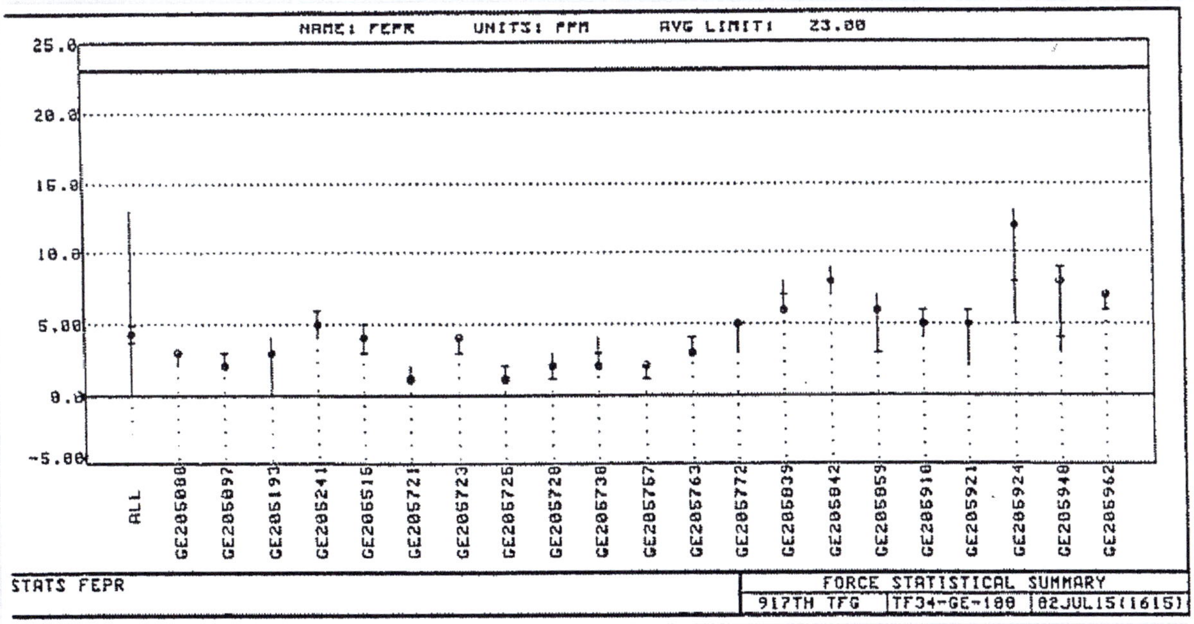

FIGURE 9

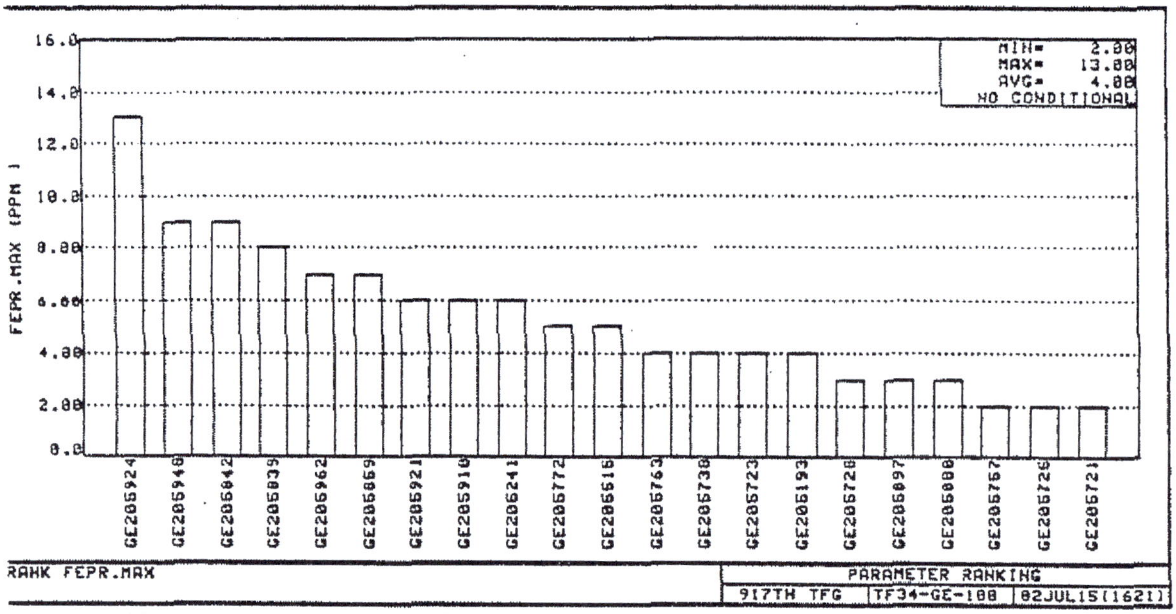

FIGURE 10

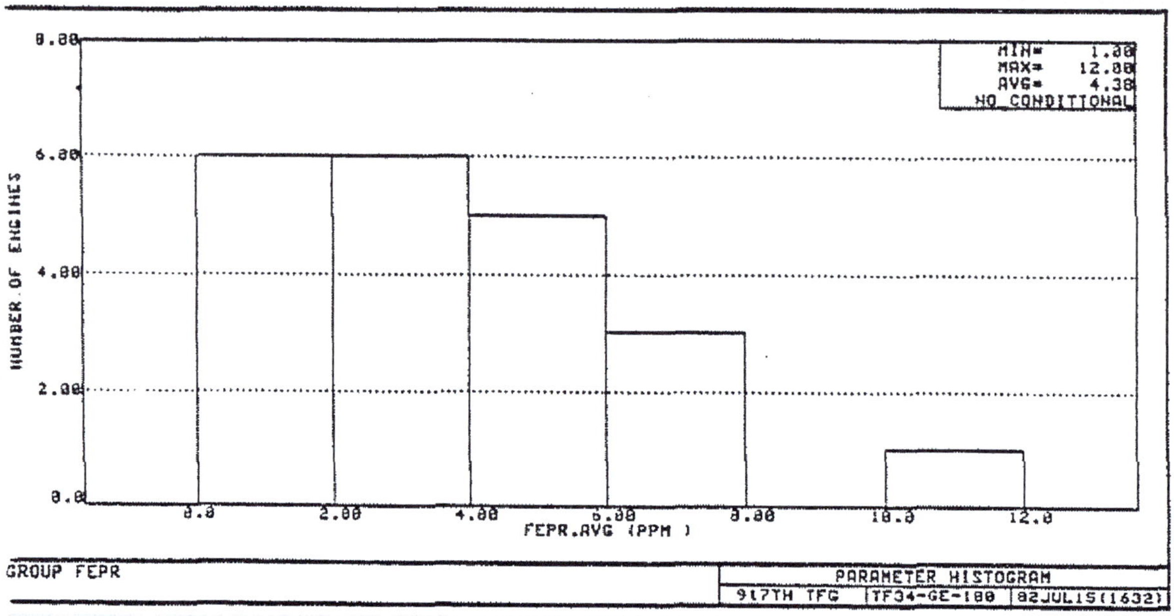

FIGURE 11

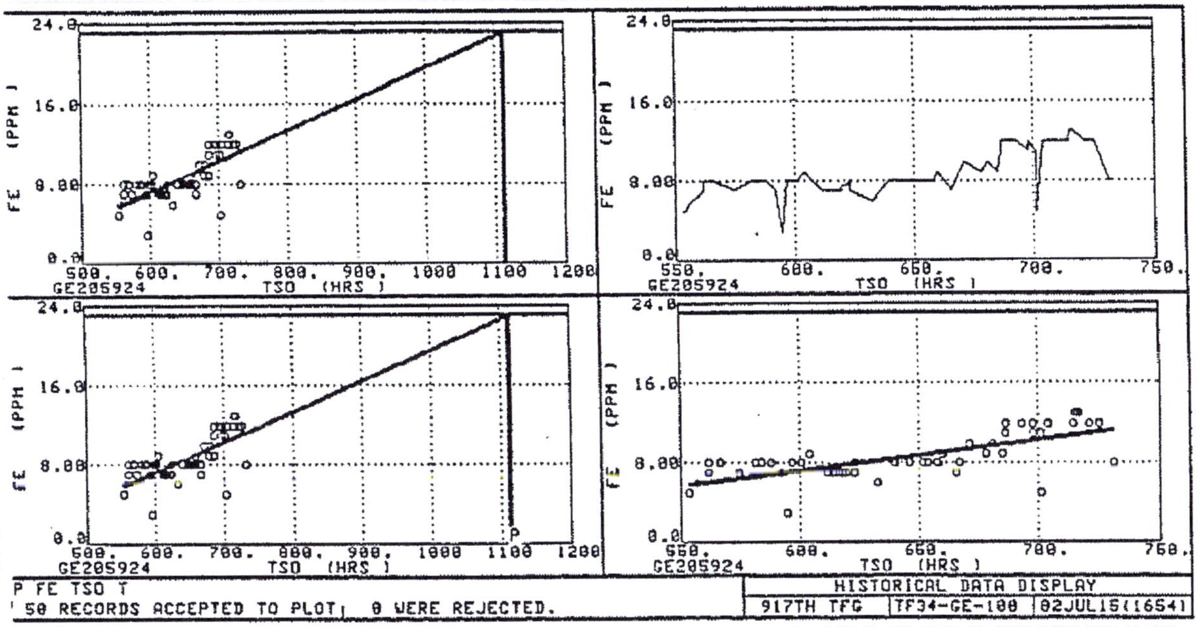

Conclusions

Performance monitoring and diagnostics have been discussed in terms of data acquisition, data processing, and maintenance information management system integration. Efficient methods for data collection and storage have been developed which should improve the accuracy and reliability of inflight data. Processing techniques which reduce this data to module directed indices show promise for predicting engine health and correlating it with engine maintenance. These tools along with other engine performance and maintenance data will be made available in a convenient, reliable, and useable format to the appropriate decision makers in the engine maintenance community. These developments should help implement RCM in the Air Force and ultimately reduce the costs associated with the Operations and Support.

Acknowledgments

This work was partially sponsored by the Air Force Wright Aeronautical Laboratories/POTC, Wright-Patterson Air Force Base, Ohio, under the contracts F33615-78-C-2062 and F33615-80-C-2047 and by the San Antonio Air Logistics Center under the contract F41608-81-C-B231.

References

1. McDonnell Aircraft Company, "Advanced Trending Analysis/EDS Data Program," AFWAL-TR-81-2125, January 1982.
2. Smish, C., Broadie, M., and De Hoff, R.L., "Turbine Engine Fault Detection and Isolation Program: Turbine Engine Performance Estimation Methods," AFWAL-TR-2058, August 1982.

CHAPTER 7

An Overview of Airborne Vibration Monitoring (AVM) Systems

R. W. Greaves
Vibro-Meter S.A

E. R. White
Flight Data Co. Ltd.

Continuous vibration monitoring of engines in flight has become routine since the introduction of FAR 25.1305 (d) (3) in 1974, not only on civil aircraft but also increasingly with military aircraft. Although early AVM systems were often unreliable, modern systems are not only reliable and credible but also cost-effective.

This paper will discuss briefly the principal technical components of AVM systems, and then describe "the current state of the art" and the uses and effectiveness of AVM.

Introduction

AIRBORNE VIBRATION MONITORING (AVM) systems were introduced during the 1960s, principally as an aid to the on-condition monitoring of aircraft gas turbine engines. At that time, on-condition monitoring was performed manually by trend monitoring [1]. For many years vibration monitoring systems did not live up to expectations due to poor system reliability and credibility. Introduction of the piezoelectric accelerometer in place of the moving coil velocity transducer and subsequent improvement of the processing and monitoring electronics reliability, including the incorporation of digital techniques, did much to enhance the credibility of the system to acceptable levels.

Credibility is not enough: AVM should also be cost-effective. A survey will be given in this paper of the many aircraft types fitted with on-board vibration monitoring systems and various case histories presented to show the usefulness and cost savings of installing AVM.

The Essential Elements of AVM

An AVM system is composed of a few key elements, namely:

- Accelerometers
- Interconnecting cable assemblies
- Processing electronics
- Displays

The "raison d'etre" of an AVM is that the health of rotating machinery is known to be dependent on its vibration. The amplitude of vibration of a rotating part at its fundamental (1/rev) frequency is a direct indication of the state of balance of that part; the better the balance, the lower the wear and the longer the life. Information on the health of a machine may also be obtained from measurement of parameters such as pressure, temperature, and debris monitoring [2]. With the exception of debris monitoring, the information content within these parameters is severely limited since they are point measurements. On the other hand, the information value contained within a vibration spectrum is far greater than for all other parameters because the measurands are amplitude and frequency of the acceleration.

Historically, the problem has been that a vibration signature contains too much information and its reduction to a comprehensible form that enables specific information to be obtained about the state of the various elements of a gas turbine is a complex task. The task is not simply that of correlating a specific frequency to a particular rotating or vibrating part but also to interpret the temporal changes in vibration amplitude and frequency. With the advent of reasonable cost and small and lightweight microelectronics, it is now feasible, even with on-board systems, to carry out sophisticated signal reduction techniques to obtain specific information from the wide frequency spectrum of a gas turbine or helicopter transmission, providing an indication of the health of its various mechanical components.

ACCELEROMETERS AND THEIR INSTALLATION—A piezoelectric accelerometer is a mechanical assembly of a number of piezoelectric forms (discs, squares, or rectangles) and a seismic mass. A piezoelectric material is one that gives an electric output when pressure is applied to it—or conversely, changes its dimensions when an electrical signal is applied to it [3].

From the basic axioms of physics that

$$\text{Pressure} = \text{Force/unit area} \tag{1}$$

and

$$\text{Force} = \text{Mass} \cdot \text{Acceleration} \tag{2}$$

and the electric output of a piezoelectric material

$$= \text{dij} \cdot \text{Applied pressure} \tag{3}$$

where dij is the piezoelectric constant of the material,
then a device may be built whose electrical output

$$= \text{dij} \cdot \text{Mass} \cdot \text{Acceleration} \tag{4}$$

and in more easily recognizable form where g is the acceleration due to gravity and the electrical unit is defined as picoCoulombs (pC), then the accelerometer sensitivity in terms of pC/g is

$$= \text{dij} \cdot \text{n} \cdot \text{m} \tag{5}$$

where

n is the number of piezoelectric forms used

m is the seismic mass

Typical examples of accelerometer constructions [4] are shown in Figures 1 and 2.

It is worth noting that the accelerometer's temperature of operation is absolutely limited by the Curie Temperature of the piezoelectric material. This is the temperature beyond which the material permanently ceases to exhibit the piezoelectric effect.

A piezoelectric accelerometer is basically a mass spring system with a resonant frequency (Fr)

$$= m \cdot \delta \tag{6}$$

where δ is the spring constant.

It is a "hard" spring system and the mechanical Q is high. The frequency response of the accelerometer with respect to frequency is flat with a high frequency and high Q resonance. This occurs in both the direct and transverse axes, whether it be a compression or shear mode device.

An accelerometer has to be mounted onto the gas turbine (or other equipment) that it is monitoring and this is usually done by the intermediary of a mounting bracket. The bracket will also exhibit a variety of mechanical resonances in the different axes within which it may experience vibration. A typical response curve of an accelerometer and its mounting bracket to an input sinusoidal vibration is shown in Figure 3. It has to be remembered that this response will factor the actual vibration spectrum of the machinery being monitored (see Figure 4) and needs to be taken into account not only when interpreting the data but particularly when designing the processing electronics.

FIGURE 1 Compression mode accelerometer.

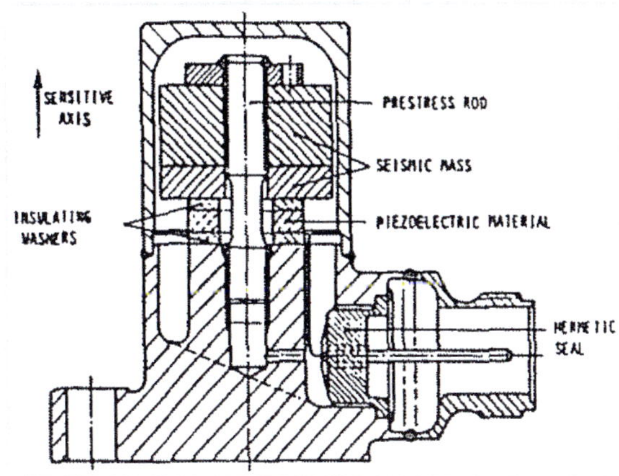

FIGURE 2 Shear mode piezoelectric sensing element.

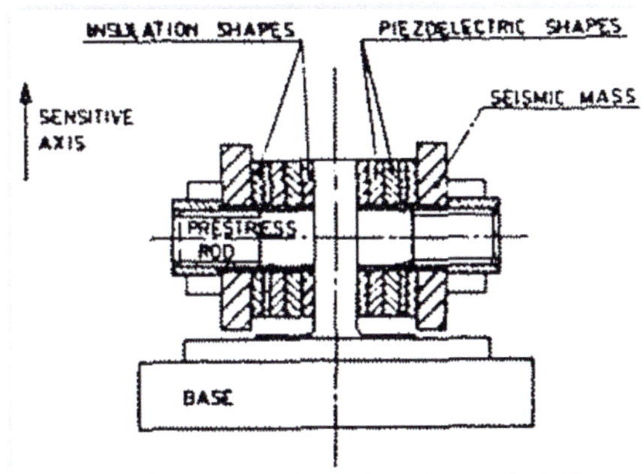

Bearing Accelerometers—It is clear that if one wishes solely to look at the unbalance of the main rotating assemblies then an accelerometer mounted onto the bearing housing will present a frequency spectrum closely related to the shaft vibration. For example, on the No. 1 bearing housing of a gas turbine, it will be sensitive to fan unbalance and relatively insensitive to low pressure turbine (LPT) unbalance. The phase of the unbalance can be clearly determined and is less susceptible to varying mechanical phase lags than it would be if it were mounted on the engine carcass. Mean Time Between Failure (MTBF) figures for fan bearing accelerometers in excess of 200,000 h are currently being obtained so the fan bearing mounting position has become standardized on the civil engines of some engine manufacturers. Historically, the reason that the implantation of fan bearing accelerometers has been keenly promoted by the airlines is due firstly to the enormous cost savings that may be obtained by constantly running an engine with the fan in a good state of equilibrium, and secondly, on more recent engines, the incorporation of a 1/rev pulse into the engine tachometer signal has enabled the on-board AVM to compute fan trim-balance data during revenue flight, which eliminates the need for expensive, time-consuming ground runs.

The siting of the accelerometer has been found to be vital to the eventual success of the AVM, and helicopter transmission gearboxes, with their large number of gears and

FIGURE 3 Accelerometer and mounting bracket frequency response.

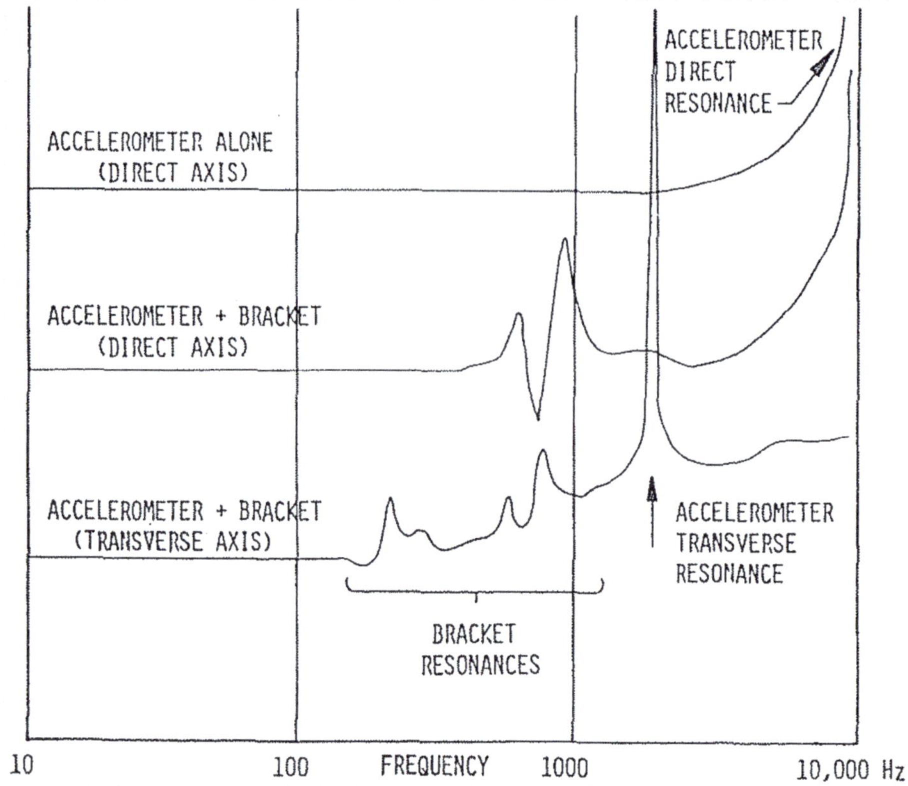

FIGURE 4 Typical engine vibration spectrum.

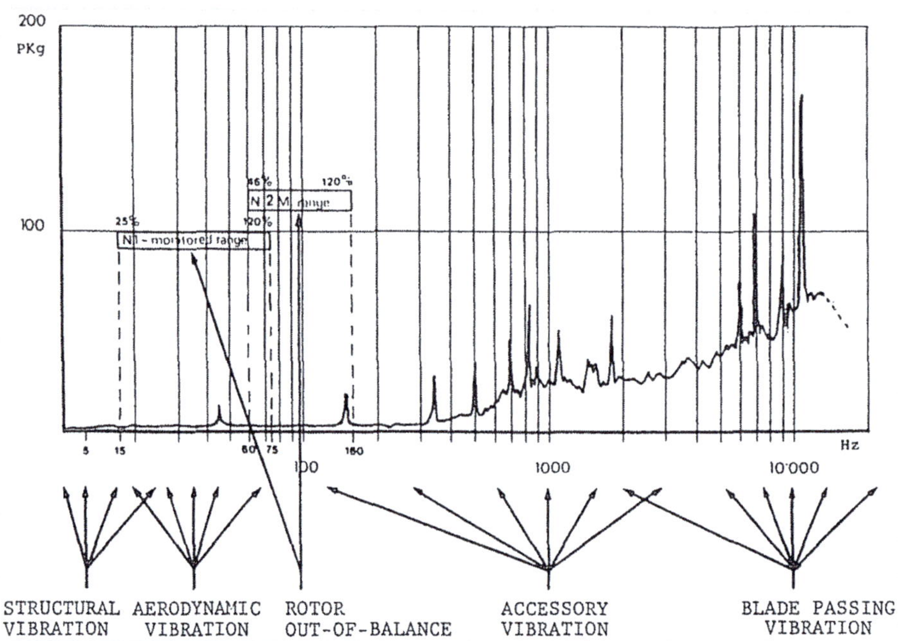

bearings, present a particular challenge here, if all rotating parts are to be successfully monitored.

Dual element accelerometers—Service experience clearly shows that the most reliable component of an AVM system, having very high MTBF figures, is the accelerometer largely due to the absence of moving parts. However, the signal transmission path, cables, and connectors are subject to wear and tear and misguided maintenance practices. Further, the electronics processing unit has a relatively low MTBF, 30,000 h being typical of today's modern systems.

In order to enhance the credibility of the data from the AVM system, there has been a move toward dual accelerometers. That is to say there are two piezoelectric sensing elements incorporated within one housing, each with its own connector. Each sensing element is cabled separately to the electronics processing unit, which continuously compares the output signal from one measuring chain with that of the other, and if the outputs should differ by more than a predetermined amount, then the system is declared inoperative.

CONNECTORS AND CABLES—A variety of connectors are used by the AVM industry, the most preferred with accelerometer manufacturers being a 7/16″-27UNS-2A, two pin connector, but current civil engines tend to use the MIL-C-83723 standard and new military engines the MIL-C-38999.

It must not be forgotten that the accelerometer output is a high impedance electrical charge and therefore the cable up to the charge amplifier must be of the low-noise variety incorporating antimicrophonic layers. PTFE insulated cables can be used to 260°C (500°F), but their wear resistance is not good. Recently progress has been made by incorporating a variety of Viton and steel overbraids to enhance mechanical performance and electrical shielding. Once the accelerometer signal has been converted to low impedance, normal shielded twisted pair cable may be used. For high temperature installations, hermetically sealed, mineral-insulated cable is used, but it should be braided not only for better wear protection but also to dampen any cable resonances between clamping points which may transmit spurious signals back to the accelerometer.

PROCESSING ELECTRONICS—A variety of signal processing techniques are in use today, ranging from wide band, tracking, and comb filters to FFT spectrum analysis. The output signal is usually presented in terms of a D.C. velocity level, thereby necessitating only one stage of integration and mean value or RMS rectification.

A typical electronics measurement chain is shown schematically in Figure 5.

More modern systems are already installed in some aircraft, incorporating microprocessor control and memory for storage of in-flight data for diagnostic and maintenance use. Current AVM systems fitted on civil aircraft contain all the processing electronics in a separate box which is installed in the aircraft electronics bay, whereas military aircraft incorporate all or part of the processing chain either into a Parameter Multiplexer (PMUX) or a Full Authority Digital Engine Control (FADEC) or separate Engine Monitoring System (EMS). If there should be space constraints within the FADEC or a high EMI environment, then a solution which has been quite successful is to build the charge amplifier, integrator, and EMI suppression into a small unit mechanically integrated with the accelerometer and to carry out diagnostic filtering within the FADEC.

The value of the data derived from AVM concerning the state of an engine is directly proportional to the sophistication of the filtering technique. For instance, a simple broad band filter covering the speed range of the engine shafts is almost useless. A vibration exceedence occurring within such a system may well be related to rotor unbalance or other relevant phenomena but may equally well be due to an accessory and irrelevant vibration. This hardly follows the requirement of the FAR to provide "an indicator to

FIGURE 5 Typical narrow-band tracking processing chain.

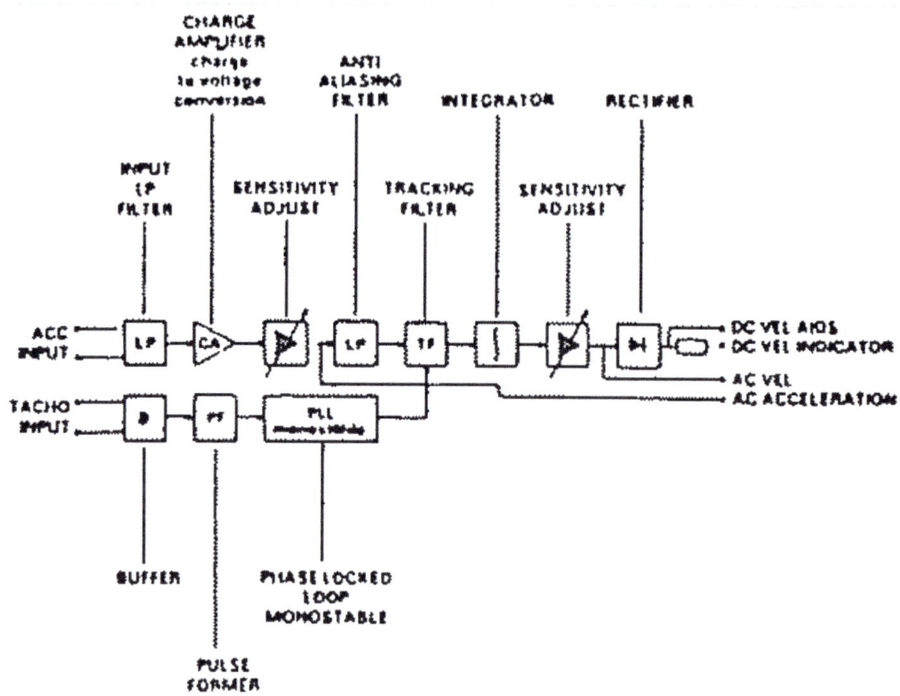

indicate rotor system unbalance." On the other hand, use of a tracking filter in which a narrow band filter is locked onto the rotor speed using the engine tachometer signal clearly does provide a system which monitors rotor unbalance [5]. In this way, the amplitude of the vibration signal due to the unbalance of the rotor is accurately determined and, if a phase reference signal (a 1/rev. signal) is incorporated in the tachometer output, then the phase of the unbalance with respect to a known reference can be determined. Storage of vibration amplitude and phase data at a variety of in-flight engine conditions will later enable ground staff to carry out a trim to balance the fan without a ground engine run [6]. Tracking filter systems are designed to specifically exclude vibration signals from all other sources. A combination of a tracking filter with a broad band filter is often felt to be optimal for in-flight engine monitoring so that fixed and variable vibrations are monitored.

Higher frequency analysis of the vibration signal is rarely carried out on board, but recording of a very broad band signal (up to 5 kHz or sometimes higher) will enable sophisticated frequency analysis to be subsequently performed to identify specific problems. Some microprocessor-based systems are capable of performing such analyses on board and may either store the results for ground evaluation or give real-time "knowledge-based" indications of vibration exceedence to the flight crew. This aspect of AVM is particularly true of helicopter transmissions, because of the need to identify incipient failures in the whole range of gears and bearings. The electronics of such a system has to select, from the whole recorded vibration spectrum, the vibration due to each component and present them in a meaningful way to the user. Techniques for doing this, as an incipient failure detection tool, have been successfully developed and are in production for health monitoring purposes.

DISPLAYS—A variety of cockpit displays are currently in use, including a warning lamp, an analog indicator, and a display on the pilot's Multi-Function Display Unit (MFDU).

The definition of the display tends to be under the control of the airframers and varies between different manufacturers.

The physical relationship between the measured vibration (in engineering units) and engine condition is complex. It is therefore difficult to present clear and credible information to flight crew (even in the form of “go, no-go”) without more knowledge-based post-processing of the data than has been present in older technology systems.

Modern technology systems “normalize” the data, thus displaying vibration margin in nondimensional units for all engine operating conditions. This is an important improvement in that it displays more meaningful information to the flight deck.

Current Use of AVM Systems

SURVEY OF ENGINES/AIRCRAFT FITTED WITH AVM—The table in Figure 6 is from an unpublished personal survey (RWG) of fixed-wing aircraft on which AVM systems are currently installed. This survey does not claim to be exhaustive, but it can be readily seen that the majority of current civil and military aircraft are equipped with AVM. It is also clear from this table that there is no consensus regarding the type of AVM to be used. However, further analysis of the survey data did reveal that the large majority of the AVM systems had similar features such as accelerometer sensitivity at

FIGURE 6 Table of engines/airframes fitted with AVM systems.

TABLE OF ENGINES/AIRFRAMES FITTED WITH AVM SYSTEMS

ENGINE	AIRFRAMES	NO. OF ACCEL.	POSITION/TYPE		SIGNAL PROCESSING	COMMENTS
CF6-50	KC-10, DC10, A300, B747, E4	2	Fan Bearing	Turbine	Broad band (Order Band)	
CF6-80	B747, B767, A310	3	Fan Bearing/Fan Frame Alt.	Turbine	Tracking Filter & Fan Trim Balance	
CFM56-2	KC135R,DC-8	1	Fan Bearing		----------	Ground Test
CFM56-3	B737	2	Fan Bearing	Turbine	Tracking Filter & Fan Trim Balance	
CFM56-5	A320, A340	2	Fan Bearing	Turbine	Tracking Filter & Fan Trim Balance	
Pegasus (F402)	AV8B	1	External, integral electronics		Comb filter	EMS
F405	T45	1	External, integral electronics		Broad Band	ADRS
TF34	A10	3	Compressor/Gearbox	Turbine	Broad Band	TEMS
CF34	CL601	1	Compressor		Order band	
TF39	C5A, C5B	2	Compressor	Turbine	Tracking filter	
TF41	A7	2	Compressor	Turbine	Broad Band/Octave Filter	IECMS
ALF 502	BAe 146, CL600	1	Compressor		Broad Band	
F404	F18, A6F	1	Compressor		Broad Band	Cable Filter
F101	B1B	2	Compressor	Turbine	Tracking Filter	
JT9D	B747, A300/DC-10	2	Compressor	Turbine	Broad Band/Tracking Filter	
JT9D7R4	B767, (A310), B747	1 (2)	Fan Frame (Fan Bearing)		Broad Band/Tracking Filter	
PW2037	B757	1	Compressor, Dual Element		Broad Band	
PW4000	B767 (A310)	1 (2)	Fan Frame (Fan Bearing)		Broad Band (Tracking Filter)	
RB211-22	L1011	2	Compressor Case, 2 each, side by side		Switched Broad Band (3 Bands)	Redundancy
RB211-524	L1011-500, MK 1, MK 2 B747	2	Compressor Case, 2 each, side by side		Tracking Filter with Broad Band	Redundancy
RB211-535	B757	1	Compressor, Dual Element		Tracking Filter with Broad Band	
RB199	TORNADO	2	Compressor	#5 Bearing	Broad Band	
M53 Mirage	F2000/F4000	1	#2 Bearing		----------	Ground Test
TAY (RB183)	GIV,(FOKKER 100)	2 (1)	2ea, side by side (Dual Element)		Tracking Filter	Redundancy
Olympus	Concorde	2	Compressor	Turbine	Order Band	
T406	V-22	3	Compressor, Turbine, Gear box		Tracking/Narrow Bands	
V2500	A320	1	Compressor, Dual Element		Tracking Filter & Fan Trim Balance FFT-Reconstituted Broad Band	

FIGURE 7 W30 VAS system overview.

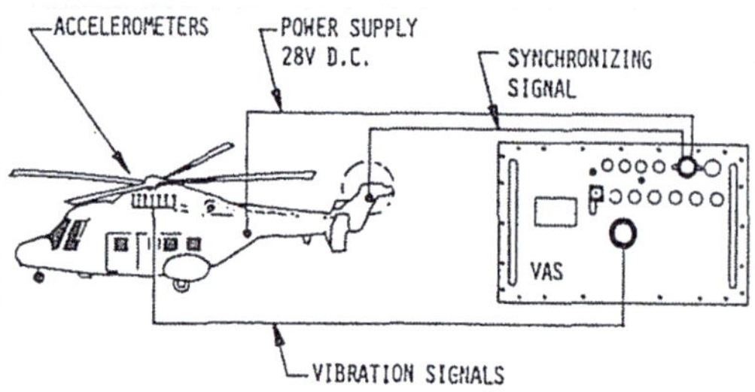

50 pC/g, two accelerometers per engine, MIL specification connectors, mechanically protected cables, and tracking filters. Helicopter applications of AVM are not yet in service in sufficient numbers for such a survey to be meaningful, but for comparison, the W30 Helicopter Vibration Analysis System (VAS) uses seven 10 pC/g accelerometers and a 3 db bandwidth of 10 Hz to 70 kHz. From a data scan of 100 revolutions of each gear being monitored, taken 12 times, the unwanted frequencies are removed by FFT and the resulting data presented for ground-based health monitoring analysis. Figure 7 shows an overview of this system. The UK Working Group on Helicopter Health Monitoring addressed all relevant techniques, of which vibration analysis was one. Their report [7] details their findings. Publications concerning the cost-effectiveness of monitoring of gas turbines and vibration monitoring in particular are rare, but the report from South African Airways [8] does conclude that monitoring is worthwhile.

Diagnostic Uses of an AVM

Which Engine? It may seem unnecessary to ask this question, but with the advent of modern aircraft construction, providing good isolation of the engines from the cockpit, the pilot can no longer "feel" which engine is vibrating excessively, and many instances have been recorded of a pilot shutting down the wrong engine on aircraft not fitted with AVM-for twin-engined aircraft this may be serious! Hence for multi-engined aircraft, a basic use of AVM is to inform the pilot which engine is vibrating excessively.

Which Part of Which Engine and When? It is of little economic benefit to inform an aircraft operator that a specific engine has attained some predetermined level of vibration. This is little better than the pilot report that "#3 is rough, I can feel it in the throttle."

Real monetary savings can only be achieved when the AVM system (or AVM plus ground database) can inform maintenance personnel (not flight crew) that a specific part, or module, of the engine has exceeded safe limits or will probably exceed such limits within the near future.

This level of diagnostic and prognostic ability is possible within modern AVM systems which have databases which may be updated enabling them to employ rule-based criteria to determine cause and effect. The perfection of such systems requires close cooperation between engine/airframe/system manufacturers and end users.

French Air Force—Laviec [9] has provided a summary of the program launched by SNECMA for the French Air Force in the early 1970s to correlate the vibration signature with the mechanical state on both the ATAR 09C and 09K50 engines.

When the system was established, the broad band vibration signature was recorded during a ground maintenance run which was:

Idle for 10 min

Accelerated to 100% for 5 s

Run at 110% for 15 s

A "normal" vibration signature was established by examining the distribution over a large number of engines of the amplitude of a range of particular frequencies. This "normal" signature then served as a reference from which exceedences in vibration amplitude could be correlated to the state of various engine parts at characteristic discrete frequencies. These frequencies were either calculated "a priori" or were determined empirically from an "experience" base. Figure 8 shows the location of engine defects that correlated with abnormal vibration signatures.

FIGURE 8 Location of correlated defects from engine vibration analysis (ATAR).

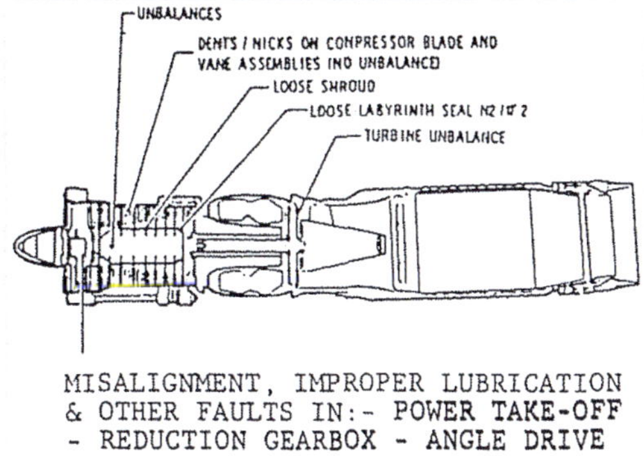

The program was started in 1975 for the ATAR 09C and in 1976 for the ATAR 09K50, and by 1978 this program was yielding extremely good results and high benefits. For example in 1978, 276 vibration analyses were carried out, and of the maintenance actions taken, the vibration signature correctly identified the defective part in 84% of the cases.

NATO Tornado Aircraft—Data of vibration-related incidents on the RB199 engine are available for the first 4 years after entry into service [10]. Of 211 reported incidents of high vibration, 72 were confirmed to be engine related, the remainder were equipment related. Of the confirmed problems, 15% were related to foreign object damage (FOD), 25% to the H.P. turbine, 55% to the I.P. turbine, and 5% to the L.P. turbine. Further investigation showed the remaining incidents to be due to poor AVM system troubleshooting and inadequate equipment maintenance.

Remedial action has drastically improved the situation.

UK Royal Air Force TriStar—These aircraft, acquired from British Airways, have been in service with the RAF under two years. However, in that time two engines have exhibited high and increasing vibration levels, which have been identified with the fixed filter band AVM to a specific spool. One engine has been removed. The health vibration monitoring trends were used to choose the optimal moment for this removal, bearing in mind the conflicting considerations of operational requirements, aircrew confidence in the engine, and the economics of maximum on-wing life versus repair costs. In this case, the presence of an adequate AVM, linked to time-series trend plots, allowed a potentially dramatic situation to be rationally handled, avoiding both the high repair costs of a failed engine and the loss of operational life due to an excessively early removal.

The second engine with high vibration is, at the time of writing, being closely monitored and will be handled in the same way.

This AVM system is used by the aircrew as a failure indication, based on the cockpit gauge display in velocity from each, or an average of both, of the two co-located transducers, and each of the three specific rotor frequency bands or a broad band. The ground-based health monitoring analysis program uses the individual rotor frequency band parameters and trends the highest peak found in takeoff or climb and a cruising flight level. It also notes exceedences of alert thresholds in any phase of flight.

Swissair A310/JT9D-7R4—On July 22, 1986, Aircraft HB-IPE showed an increase in N2 vibration on engine number two to approximately 2 in./s [11]. On this indication alone, the engine was sent for shop maintenance and the H.P. turbine shaft was found to have cracked around 150°. In general it is difficult to estimate the cost savings of an AVM, but in this case it is clearly very high (an engine could have been destroyed).

That AVM systems are highly specific (i.e., almost no false positives) is shown by the experience of Swissair. Since its introduction on the A310, over 65,000 flight hours have been accumulated to May 1987, and of all five cases where engines developed abnormal vibration readings, a significant engine problem was confirmed by shop maintenance. In three of these cases, vibration was the only indication of an engine defect.

Lufthansa A310/CF6-80—Lufthansa [12] have estimated that they are able to save a minimum of 20 "hot" trim balance runs per year by using data from the on-board AVM system to "cold" trim the fan on the ground.

FIGURE 9 Lufthansa results with the A310/CF6-80 fleet showing savings due to use of trim-balance data from the on-board AVM.

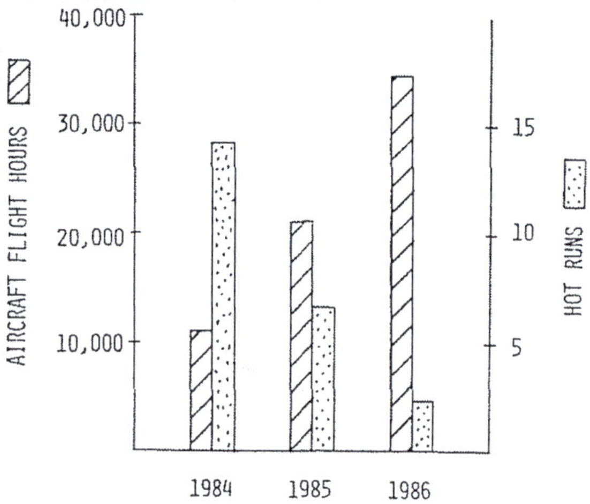

The saving per run is:

2300 kg of fuel
2 h of aircraft availability
5 to 6 man-hours

Figure 9 shows the aircraft hours flown per year and the number of "hot" trim balance runs carried out annually. The latter substantially decreases with time as the on-board data are increasingly used following the introduction of "cold" trim balancing in 1985, despite a threefold increase in flight hours.

It was also stated that the average cruise engine vibration level following an on-ground "hot" trim balance was 0.5 in./s but that "cold" trim balancing gave a level of less than 0.2 in./s. This resulted in greater passenger comfort, less pilot fatigue, less engine wear, and less accessory wear.

British Airways (BA)—AVM systems are fitted to most fleets. Boeing 737 (P&W JT8D), Boeing 747-100 (JT9D), and BAC 1-11 (RR Spey) are exceptions. The reason for these exceptions are that the engines have reached maturity and they exhibit an extremely low incidence of vibration-related failure modes. The Spey was delivered with an AVM and many cases of HP turbine failure were picked up in the incipient stages in the early years of BAC 1-11 operation. However, this failure mode was subsequently modified out of the engine and the AVM system, which employed moving-mass seismic transducers, needed constant maintenance attention to keep it in working order. The decision was then taken to disable the AVM. The JT9D vibration system, which is nearly 20 years old technology, became increasingly unreliable as the engine reliability improved. A decision to remove the AVM system was taken in 1984.

There have been literally hundreds of cases where AVM information has been used in BA, mainly to remove engines with incipient failures. Failure modes have included partially missing turbine blades, excessive blade tip rubs and spacer ring frettage, bearing failures, misplaced or missing blade locking plates, fan blade clapper lockup, oil migration into compressor drums, shaft coupling out of alignment, compressor stack bolt loosening and disc frettage, etc.

Some recent case histories include the fan shaft location bearing failures on RB211, which was potentially disastrous for both engine and aircraft. This failure mode had a very rapid progression to failure, but the AVM system was able to give some seconds warning, allowing a reduction of power and preventing catastrophic failure. Another case is the sixth stage compressor failure on the RB211, which is identified by AVM early in the progression of the failure mechanism, saving some secondary damage.

The V22 Osprey Vibration Monitoring System—Task Order T-2-126 was generated by the Defense Science Board "Study on Operational Readiness" in 1981. The first report of this study [13] demonstrated that the monitoring of mechanical system condition, especially in the turbine engine (and gearbox area), could have a high payoff potential.

The V22 tilt rotor aircraft will incorporate one of the most complete monitoring and diagnostic systems ever to be produced. This system makes extensive use of accelerometers together with strain gages, temperature sensors, and other inputs to an on-board processor which monitors vibration, structural life parameters, and engine diagnostics.

The vibration section uses some 19 accelerometer inputs to monitor engines, rotors, and interconnecting drive shafts for exceedences and trend data by means of both broad band and multiple-order analysis.

In-flight track and balance data for the two prop rotors is also calculated and balance maintenance instructions are produced. An interesting feature is the use of combined acceleration and temperature sensors on the interconnecting driveshaft bearings to provide coherent trend and exceedence data.

Structural life monitoring includes aircraft loading and flight parameters combined with in-flight processing and selection of significant data.

The engine diagnostics package measures and computes power, exceedence thresholds, life-usage, performance, and surge data as well as monitoring the oil system, and these data can be correlated to specific engine/module identification parameters and the problem identified.

The Vibration monitoring, Structural Life monitoring, and Engine Diagnostics (VSLED) system is linked to the central mission computer, on-board recorder, and a data transfer unit which assures the essential link with ground-based computer systems for subsequent further analysis, correlation with other inputs, and updating of the on-board system database.

This is a prime example of the full functional integration of an on-board monitoring/ diagnostic system into a complex airplane and ground system.

Summary

1. Vibration is one of the most important indicators of engine health.
2. Piezoelectric AVM systems in current use are reliable and credible indicators of engine vibration.
3. The majority of civil and military aircraft are fitted with AVM.
4. Installing and running AVM systems is cost-effective, for example, in reducing the cost of fan trim balance and preventing secondary engine damage in the event of FOD.
5. AVM is a major contributor to flight safety [14].

References

1. Bart, H.V.J. and Nadler, A.B., "Maximum Installed Engine Utilization Through Disciplined Health Monitoring," SAE Technical Paper 750612, 1975, doi:10.4271/750612.
2. Hurry, M.F. and Holmes, M., "Military Engine Usage Monitoring Developments in the United Kingdom," *ASME Gas Turbine Conference*, London, April 9-13, 1978.
3. Cady, W.G., *Piezoelectricity*, 2 Volumes, (Dover, 1964).
4. Greaves, R.W., "Piezoelectric Transducers," *Transducer 78 Conference*, London, June 26, 1978.
5. Rubel, H.J., Peck, K.I., and McInturff, J.A., "Engine Vibration in Flight," *AIAA Aircraft Systems and Technology Conference*, Los Angeles, August 21-23, 1978.
6. Neese, W.A., "Use of Expanded AIDS in Engine Health Monitoring on the CF6-80 Engine for the A310 Airbus," SAE Technical Paper 841505, 1984, doi:10.4271/ 841505.
7. CAA Paper 85012, "Report of the Working Group on Helicopter Health Monitoring," London, August 1985.

8. Kruger, T.C.B., "Reducing Engine Operating and Maintenance Costs Through the Effective Use of Monitoring Techniques," *South African Airways, Engine Monitoring Symposium*, Johannesburg, April 19-21, 1983.
9. Laviec, G., "Le Suivi du Comportement Vibratoire des Moteurs ATAR," SNECMA, Product Support Department, 1982.
10. RB199 Product Support Document PS/JCA/MEM/SMH, dated July 9, 1982.
11. Communication from Swissair to R.W. Greaves, July 20, 1987.
12. Communication from Lufthansa to R.W. Greaves, March 6, 1987.
13. Howard, P.L., OSD/1DA, RSM Study, "Mechanical Systems Condition Monitoring Working Group Final Report," 1986.
14. SAE E-32 Committee, "Lessons Learned From Developmental and Operational Turbine Engine Monitoring Systems," SAE AIR 1871A.

CHAPTER 8

Advances in Commercial Engine Lube Debris Monitoring

Paul D. Higgins and John T. Crow
Vickers Inc.

Wear debris monitoring is the most effective technique available for detecting impending failure of oil-wetted components. The majority of the gas turbines operated in a flight environment utilize magnetic chip collector (mag plug) inspections to detect the onset of oil-wetted component failures. However, successful chip collector usage depends upon timely inspection and the subsequent disposition of recovered debris. Shortcomings in these critical maintenance actions, and/or wear modes which result in product failure between scheduled inspection intervals, compromise the effectiveness of this method of wear debris monitoring.

The General Electric (GE) GE90 engine utilizes the Quantitative Debris Monitor (QDM®), developed by Vickers, Tedeco Division. The QDM system—the first and only advanced, remote-indicating lube debris monitor qualified for commercial flight service—continuously monitors the lube oil for the presence of wear debris. The system is capable of counting individual debris particles and distinguishing their size relative to a preset "target" threshold. These advanced features permit "on-condition" maintenance, and distinguish QDM from mag plugs and electric chip detectors, which may be subject to costly "nuisance" debris indications.

Performance and qualification requirements were developed in cooperation with GE and its French partner, SNECMA. The QDM system was qualified to customer specifications and released as standard equipment on initial GE90 entry into service. Field experience since service release has confirmed the system's ability to reliably and accurately indicate impending oil-wetted component failure well in advance of the loss of functionality.

Wear Debris Monitoring Fundamentals

The capture and analysis of magnetic, metallic debris for the purpose of detecting impending failure of oil-wetted parts in helicopter gearboxes began to gain widespread acceptance in the 1950s. Much of the technology was developed in cooperation with military users. Since turbine engine gears and rolling element bearings exhibit the same failure modes as gearboxes, lube oil debris monitoring has been applied to nearly every military and commercial flight engine now in service. No other type of engine monitoring system can provide such positive proof of impending failure.

Surface contact stresses result in the spalling and pitting of load path surfaces and ultimately in the failure of gears and rolling element bearings. Circumstances causing spalling and pitting include corrosion and bearing separator failure, which can cause a change in load path. The debris separating from these surfaces becomes entrained in the lube oil and can be captured on magnetic chip collectors strategically placed in the lube system. These "mag plugs" may be removed at regular intervals and examined for the presence of such debris.

The capture and retention of wear debris is of critical importance, since subsequent examination and positive identification of the debris is necessary for the operator to make conclusive maintenance decisions. Most engine manufacturers make recommendations for maintenance or removal from service based on the type, size, and source of wear debris recovered from magnetic chip collectors. In addition, operators build an effective knowledge base of the engine types they operate by collecting, identifying, and saving wear debris samples from fielded engines. Many also correlate wear debris finds with worn components removed from engines at overhaul. Since military and major commercial fleets operate dozens or even hundreds of the same type of engines, maintaining an in-house expert (or staff) along with a logistics system to analyze and track wear debris and make appropriate maintenance decisions is easily justified.

Application of Debris Monitoring to Flight Engine Maintenance

Military flight engine operators have typically had the time and personnel to perform maintenance, including inspection of chip collectors, after every flight or at frequent intervals. Commercial operators have the same opportunity but depend upon a careful balance of cost (time) and flight safety. Chip collectors are typically inspected during scheduled Letter Checks. Lube oil debris finds are sent to an in-house lab for positive identification of material type; however, positive identification of the debris may not be possible before the aircraft must be dispatched again. As a result, field maintenance personnel are often equipped with samples of previously collected/identified wear debris for comparison with newly collected debris finds. The amount of debris collected during an inspection strongly influences the decision regarding continued operation. Sometimes engines are put on a "maintenance watch" and the chip collector is inspected more frequently. If a subsequent increase in the rate of generation of wear debris is noted and/or bearing material is positively identified by the lab, an engine will most likely be removed from service. An exception would be the opportunity to service a line-replaceable component or engine module on-wing.

Improper chip collector handling has resulted in oil loss events during engine operation due to missing, incorrect, or improperly installed O-rings or magnetic plugs. The most famous example involved an Eastern Air Lines L-1011 which experienced in-flight oil loss from all three engines' chip collectors due to a mechanic's failure to reinstall O-rings on several plugs following a routine inspection. This event prompted the introduction of more murphy-proof hardware; however, even these safer designs can be defeated by circumstances created within the framework of field maintenance. When an engine family exhibiting a chronic problem is put on a maintenance watch, the probability of collateral events related to probe inspection and handling is further multiplied. Failure modes which progress to the loss of component function between inspections are another unacceptable scenario. Such events will obviously proceed undetected, resulting in in-flight engine problems due to oil-wetted component failure.

Quantitative Debris Monitoring (QDM)

The development of an effective, engine-mounted debris monitoring system requiring only on-condition inspection provides a solution to the problems described above. Acceptance of such advanced capability systems has of course hinged upon the cooperation of engine manufacturers and operators. Vickers, Tedeco Division, designed the Quantitative Debris Monitoring (QDM®) system to meet the maintenance needs of the flight engine operator. The QDM system for the GE90 engine is the first and only such system qualified on an FAA-certified power plant. Extensive development and qualification testing ensured conformance to specification requirements critical to failure detection and proper engine lube system integration. Key system parameters include debris capture and indication efficiency (of "target" sized particles), hydraulic pressure drop, and compatible sensor signal conditioning electronics design. Target size is defined as the minimum particle mass representative of the debris population which separates from load-bearing surfaces during component failure due to surface contact stress.

Figure 1 illustrates the general layout of the QDM system within the GE90 lube system. The QDM sensor and sensor housing (Lubriclone) are installed immediately

FIGURE 1 GE90 QDM/lube system schematic.

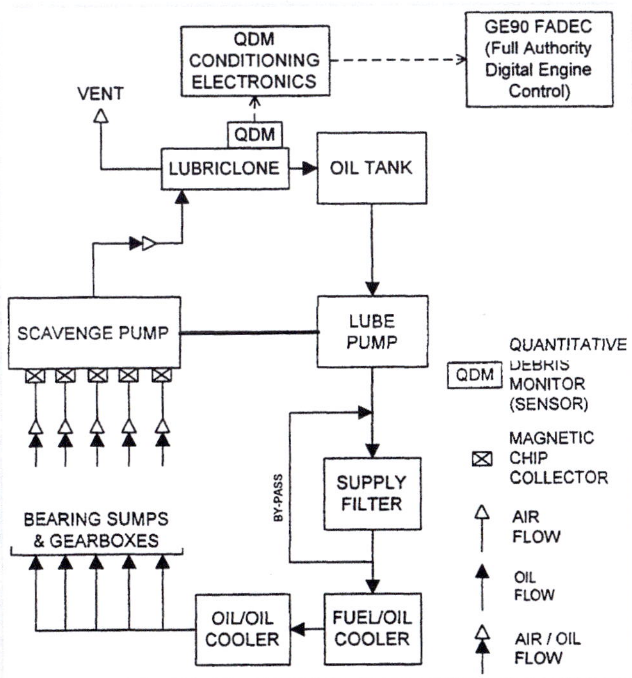

FIGURE 2 QDM system installation on GE90 oil tank and fan case.

downstream of the scavenge pump at the inlet of the oil tank, which is mounted to the engine fan case. Figure 2 shows the actual on-engine installation and relative size of the components. This location facilitates inspection of the debris sensor through the same access panel used to service the oil tank. The sensor is installed in a valve which automatically closes when the sensor is removed, allowing inspection without the loss of oil. The Lubriclone provides debris capture efficiencies of greater than 75% of target-sized debris at takeoff and cruise conditions. By definition, this means that for every one hundred target-sized particles flowing through the housing, a minimum of seventy-five are captured and retained on the sensor. By comparison, a magnetic probe simply installed through the side of a lube system scavenge pipe would typically capture 25% or less of the target-sized debris flowing by. This significant increase in debris capture efficiency is a key factor in providing earlier warning of impending failure. The hydraulic pressure drop across the Lubriclone is well within customer-specified limits.

Isolation of the source of wear debris to a particular bearing sump or gearbox is accomplished through the use of "slave" mag plugs installed in the individual scavenge lines. Since they are upstream of the "master" QDM sensor, they are engineered to collect only enough debris to isolate the failure. Once the QDM system indicates the presence of sufficient wear debris to warrant an inspection, these plugs may be removed and inspected. Fault isolation may provide the opportunity to perform maintenance on a line-replaceable component or engine module on-wing, eliminating unnecessary engine removal.

QDM vs. Electric Chip Detectors

The QDM system is capable of capturing and indicating the presence of *individual* magnetic debris particles above a customer-defined target mass. The system electronically counts these particles and produces an alarm when a preset total chip count or rate of debris generation is exceeded. Alarm thresholds are typically software configurable. These advanced features distinguish QDM technology from electric chip detectors, which indicate the mass presence of debris when a set of contacts is bridged by accumulated magnetic particles. Chip detector remote indication is normally provided by an audible alarm or cockpit/maintenance panel light.

While electric chip detectors may be applied effectively, a primary service problem exists. The amount of debris required to close the circuit can range from one large, well-placed sliver up to a hundred or more target-sized particles, depending both upon the nature of the debris population and the design of the chip detector. Further, the contacts may be bridged with *any* size of debris, regardless of its significance. Some wear debris is always present in any oil lubricated machine; however, particles which are smaller than the target size are typically not indicative of the onset of component failure. Since this normal wear debris is not of interest to the engine operator, chip detectors are prone to "nuisance" debris indications. For the operator, this represents increased workloads to remove, inspect, and interpret the debris found on the chip detector.

The QDM system, by means of electronic intelligence, ignores this fine debris, avoiding nuisance warning indications. When a bearing sheds debris due to skidding or separator contact with races, the resulting particles are typically smaller than the

FIGURE 3 Typical bearing roller and race damage.

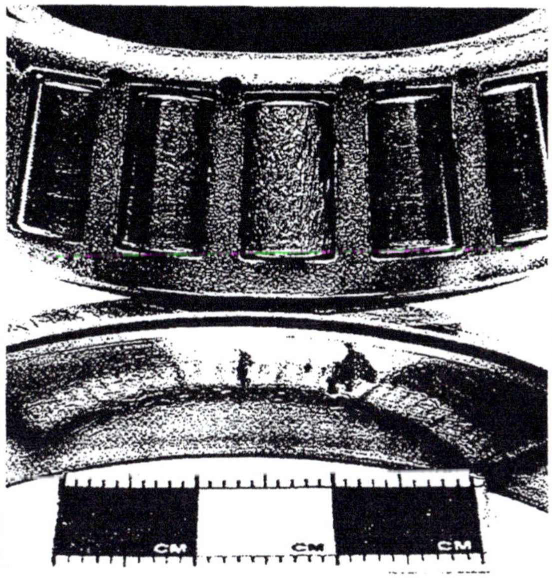

target size. However, these events are very likely to progress to a typical spalling failure due to the change in load path on the bearing. Therefore, QDM provides protection for this type of failure as well. Figure 3 illustrates a bearing with substantial roller and race damage. The QDM system would have indicated the presence of wear debris well in advance of this condition.

GE90 QDM System Development, Qualification, and Performance

GE's interest in making on-condition debris monitoring standard equipment on the GE90 engine can be traced to their positive experience with Vickers' QDM systems during development testing for the YF-120 engine program in the early 1990s.

Electromagnetic debris detection is dependent on a large number of factors, including particle mass and shape, physical orientation, and velocity at capture. The large number of variables results in a bell-curve distribution of debris voltage pulse levels for a given material/mass combination. QDM systems are engineered to maximize the probability that all debris equal to or greater than the target particle mass will be counted, while minimizing the possibility of indicating particles below the threshold. For GE90, an indication efficiency of 85% of targeted particles was specified at all oil flow conditions (Figure 4).

Since the magnetic characteristics of different steel alloys can vary significantly, it is critically important that the debris material used in performance testing match that of the engine component the system will monitor. As such, Vickers obtained samples of the specified M50NiL bearing material early on for use in development testing on its hydraulic flow stand. M50NiL is significantly less magnetic than many other commonly used bearing steels.

GE and SNECMA originally specified a "detection threshold" of 0.05 mg (approximately equivalent to a .020″ × .020″ × .001″ steel flake), and began testing the QDM

FIGURE 4 GE90 quantitative debris monitoring system components.

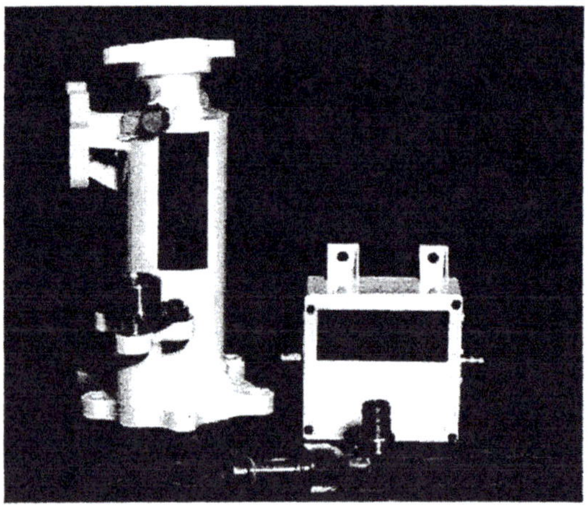

system in 1992 following delivery of initial engineering development systems. During actual on-engine development tests in 1993-1994, GE determined that a larger target size would better suit their engine maintenance goals, and the system specification was changed to 0.13 mg. The resulting design change was accomplished through an adjustment in the Signal Conditioner electronics. The QDM System was qualified in 1994 after successfully completing a broad range of environmental and performance tests, including vibration, shock, acceleration, thermal cycling, and humidity.

The GE90 was part of the first ever "ETOPS out-of-the-box" engine/airframe development program, and required ultrahigh reliability across the board. Thus, GE and SNECMA sought a monitoring system which would be capable not only of accurate, reliable oil-wetted component failure detection but also extremely resistant to nuisance debris counts. The QDM system is limited by specification to a nuisance count rate of no more than *one every million* flight hours. In hundreds of hours of on-engine utilization to date, both in GE90 development testing and on Boeing 777 revenue service, no nuisance counting has been reported. Further, the correlation between electronically recorded debris counts and the actual debris recovered from the QDM Sensor has been nearly perfect, exceeding the 85% indication efficiency requirement, and approaching 100%.

The QDM system's value was recently substantiated when a problem was discovered with one of the GE90's accessory gearbox bearings. British Airways was authorized by the FAA to check QDM debris counts on a daily basis as a means of monitoring the suspect bearing. In this "maintenance watch" situation, the governing authorities have also mandated manual inspection of the QDM sensor at very conservative, two-day intervals. To date, the correlation between recorded counts and actual debris finds has been 100%, with as many as ten target-sized chips found on the sensor at one time.

Conclusion

The QDM system is an accurate, reliable on-condition system for indicating failure of oil-wetted components in advance of the loss of component function. Coupled with magnetic chip collectors placed in the engine oil scavenge system in a "master/slave" arrangement as shown in Figure 1, QDM provides effective fault isolation. The system has proven its reliability and effectiveness in GE90 engine revenue service, promising increased flight engine safety and reliability at reduced operating cost.

CHAPTER 9

Very High-Frequency Monitoring System for Engine Gearbox and Generator Health Management

Matthew J. Watson and Carl S. Byington
Impact Technologies, LLC

Alireza Behbahani
Air Force Research Laboratory, Wright Patterson Air Force Base

In cooperation with the major propulsion engine manufacturers, the authors are developing and demonstrating a unique very high-frequency (VHF) vibration monitoring system that integrates various vibro-acoustic data with intelligent feature extraction and fault isolation algorithms to effectively assess engine gearbox and generator health. The system is capable of reporting on the early detection and progression of faults by utilizing piezoelectric, optical, and acoustic frequency measurements for improved, incipient anomaly detection. These gas turbine engine vibration monitoring technologies will address existing operation and maintenance goals for current military system and prognostics health management algorithms for advanced engines. These system features will be integrated in a state-of-the-art vibration monitoring system that will not only identify faults more confidently and at an earlier stage but also enable the prediction of the time to failure or a degraded condition worthy of maintenance action.

The authors have made significant progress toward identifying, computing, and comparing the high-frequency feature sets generated with various vibro-acoustic measurement techniques. Specifically, the technology has been demonstrated on two subscale test stands. The first is a generator test rig that was equipped with a laser vibrometer and two high-frequency accelerometers. Various mechanical and electrical faults were seeded, with an emphasis on generator bearing faults. Initial results show very good detection capability in frequency bands well above those used in traditional vibration analysis. Another focus, accessory gearbox systems, was addressed for feasibility through a gearbox test rig, which was instrumented with high-bandwidth accelerometers and wideband and narrowband acoustic emissions (AE) sensors. Baseline, seeded fault, and fault progression tests were conducted, including tests with various levels of gear tooth corrosion. Successful detection of this fault was then demonstrated using a number of new, innovative approaches. A statistical analysis was also performed to compare the approaches, with narrowband AE and high-frequency vibration features performing the best.

Introduction

The ability to successfully detect and isolate faults is critical to the performance of diagnostic algorithms and the implementation of Prognostics and Health Management (PHM). Prognostics rely on incipient (early) fault detection and isolation to provide a reliable and timely prediction. A well-designed PHM system seeks to extend, as far as practical, the feature's detection horizon. The detection horizon is the elapsed time between the first detection of a fault and the resultant mechanical failure. Figure 1 shows a timeline representation of several diagnostic feature types and their order with respect to each other in increasing detection horizon. Incorporating features that increase detection horizon is key in the design of a high-performance diagnostics/prognostics system.

Vibro-acoustic data continues to provide some of the most quantitative and reliable indicators of bearing, gear, and rotating member fatigue detection and diagnosis. The indicators are typically spread throughout the vibro-acoustic regime. Figure 2 illustrates the regions of response, health management uses, and sensing capabilities of vibro-acoustic data. Healthy machine vibration energy for a gas turbine engine dominates the frequency region from DC through 10 kHz or so. This region is also appropriate for rotordynamic fault detection, such as misalignment and imbalance. The typical utility of high-frequency measurements to the diagnostics and prognostics approach is documented in several studies [1, 2, 3]. For instance, the earliest indications of bearing

FIGURE 1 Typical turbomachinery diagnostics detection horizon comparison.

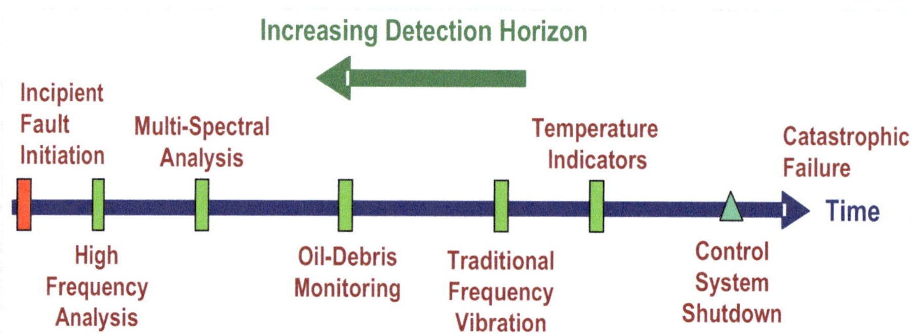

FIGURE 2 Vibro-acoustic spectrum, health management uses, and enabling sensing capability.

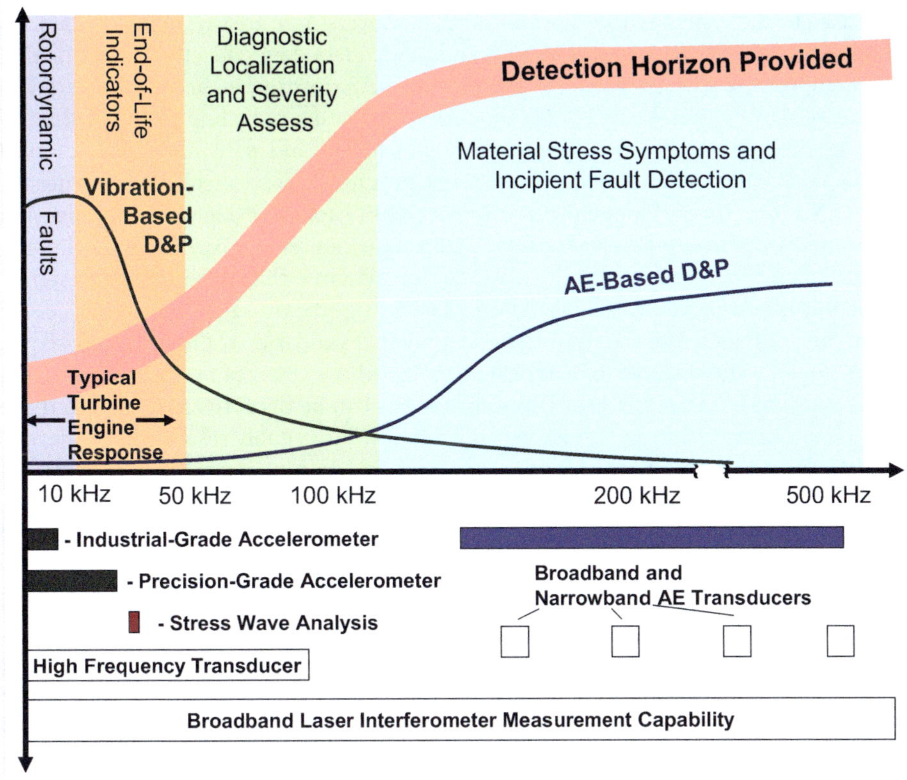

problems appear in ultrasonic frequencies (>30 kHz). As wear increases, the component noise drops in frequency range.

During fault progression, slight defects begin to ring the bearing at natural frequencies and overall high-frequency energy and demodulated spectra values increase. Further in the progression, bearing defect frequencies and harmonics appear in the conventional spectrum analysis (if the overall machinery noise is not too high). As wear progresses, more harmonics appear with stronger sidebands around the defect frequencies. High-frequency demodulation and enveloping confirms this progression of damage. At the very end of life, the magnitudes of 1 times RPM are affected and more harmonics appear in the frequency analysis. Defect frequencies also start to disappear and are replaced by high-frequency random noise as the damage induces more random, chaotic vibration. Just prior to failure, spectrum energy will usually grow by excessive amounts.

The authors have applied their in-depth vibration analysis knowledge to select and integrate an optimal set of VHF features into a comprehensive vibration monitoring system that is capable of incipient fault detection and fault progression tracking. In the end, enormous economic savings will be realized by improving overall engine diagnostic and prognostic capabilities through optimal utilization of the proposed technologies. The VHF vibration measurement and associated feature analysis, when combined with predictive models, will help reduce time to diagnosis, reduce engine removal rates, produce optimal maintenance and inspection intervals, and reduce support costs for these critical areas of concern.

FirstCheck™: Sensor Validation

An important assumption in the deployment of an automated PHM system is that the data used by the system is accurate and valid. However, there are various factors associated with sensor hardware degradation and inadequate data collection methods that can compromise the integrity of vibration data. For example, accelerometers can be damaged by exposure to excessive shock or temperature or by improper handling by maintenance personnel. Other factors are more insidious and arise from loose electrical connections, poor solder joints, loose mounts, ground loops, electromagnetic interference (EMI) and Radio Frequency Interference (RFI) noise, or degradation of sensing instrumentation due to thermal effects. Data acquisition effects, such as A/D clipping and insufficient dynamic range, can also alter the dynamic characteristics of the signal. These issues can be very problematic and lead to significant safety concerns (i.e., on board) and cost increases (i.e., during development or validation testing, where lost data means that a test may have to be repeated). In addition, changes in the dynamics of a vibration signal characteristic due to sensor faults can be deceptively similar to those due to mechanical failures (or vice versa), which will inevitably result in false alarms. Rigorous and automated analysis of the integrity of vibration data is therefore critical to providing accurate health assessments.

Based on the authors' experience, vibration monitoring algorithms can be impeded by faulty accelerometer data. Figure 3 shows the result of the authors' analysis of a gear pinion failure that occurred on the test stand of a high-speed (thousands of RPMs), high-power (tens of thousands of horsepower) military fighter aircraft drive train. As seen, several vibration features react simultaneously, indicating that a potential fault is present in the system. Information gathered solely from this sensor would confidently indicate a fault. However, upon further investigation of the raw sensor data (shown in the top plot of Figure 3), one can see that this reaction was caused by faulty (intermittent) data and, therefore, should not be trusted.

In order to address this potential source of false alarms and validate the integrity of the signal, the developed approach first evaluates the high-frequency vibration

FIGURE 3 False alarm caused by faulty sensor.

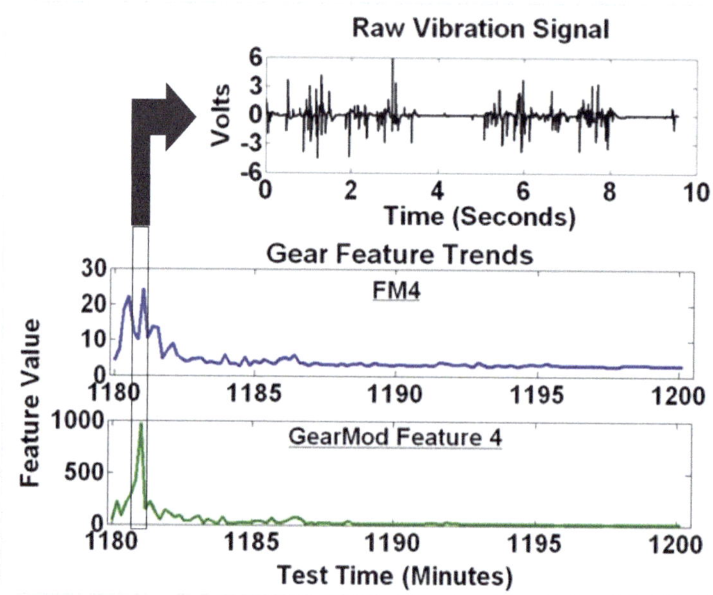

signal using a technique termed FirstCheck™. This technique tracks specific signal characteristics and statistical-based features to identify basic sensor failures, such as clipping, weak signal, overamplification, and DC bias, as well as other forms of corrupt data. This approach is more effective than traditional energy measures (i.e., peak-to-peak strength), which cannot detect a corrupt vibration signal when its values are within normal range but lacking in frequency content. Similar to mechanical fault detection algorithms, the developed approach uses a baseline of healthy sensor values to ensure that the algorithm does not disregard a valid signal.

ImpactEnergy™: Bearing Fault Detection and Isolation

Within the developed approach, bearing fault detection and isolation is performed using a set of algorithms termed ImpactEnergy™. Although bearing characteristic frequencies are easily calculated, they are not always easily detected by conventional frequency domain analysis. Incipient bearing damage is most often characterized by short-burst impulses in the vibration signature. Vibration amplitudes at these frequencies due to incipient faults (and sometimes more developed faults) are often indistinguishable from background noise or obscured by much higher amplitude vibration from other sources, including engine rotors, blade passing, and gear meshing. Similarly, time domain energy features, such as RMS and Kurtosis, are not significantly affected by such short bursts of low-intensity vibrations. Therefore, traditional time domain or frequency domain analyses often encounter problems in detecting the early stages of bearing failure.

The developed algorithms integrate traditional time-domain statistical analysis and frequency-based spectral analysis techniques with high-frequency demodulation and advanced feature extraction algorithms to provide a more effective PHM solution. The advantages of using the high-frequency response to identify and track bearing damage are well documented [4, 5] and proven to be effective. Demodulation (or enveloping) allows the broadband energy caused by failure effects to be differentiated from normal system noise. This approach provides the ability to detect defect impulse events much easier than traditional analysis techniques. A key consideration is selecting the bandpass filter that is centered on the expected carrier frequencies. Through proprietary knowledge and field-application experience, the authors have developed a process to identify key carrier frequencies [6, 7].

For complete characterization of bearing health from incipient fault to failure, the algorithms include processing to extract an extensive set of time and frequency domain features from both the raw (unprocessed) and processed vibration signals. This extensive feature set provides an effective fault isolation capability. Time domain features include traditional statistical measures, such as RMS, Kurtosis, and Crest Factor. Frequency domain features include the power levels of specific bearing defect frequencies, which are compared against known, healthy baseline thresholds. These features can be very useful in diagnosing a fault [7]. In addition, observing the magnitude of the rate of change of these features can also provide a prognostic benefit.

GearMod™: Gear Fault Detection and Isolation

The authors have developed a set of algorithms, termed GearMod™, that are used to extract diagnostic features that can be employed for gear fault detection and isolation. These algorithms contain a broad range of statistical methods based on time synchronous averaged (TSA) and other processed signals. The time synchronous averaging technique is a very useful noise reduction tool that reduces random noise levels and disturbances from events unrelated to the gear of interest. TSA has been extensively used to preprocess gear vibration signals [8, 9]. The fundamental principal of TSA is that the vibration signals related to shaft and gear rotation will repeat periodically with each rotation. Therefore, TSA divides the vibration signal into contiguous segments (with each segment representing one shaft rotation) and calculates the average of the segments. This process reinforces vibration components that are synchronous to the shaft rotation and cancels out others that are out of phase in consecutive rotations. The algorithms calculate time-domain features, such as RMS, Skewness, Kurtosis, and Crest Factor, as well as features from the spectrum of the averaged signal. Additional features are also calculated using proprietary methods [10, 11].

Statistical Analysis and Threshold Setting

Statistical detection analysis, for the purposes of selecting and implementing an optimal threshold for "calling out" specific fault or anomalous conditions, is based upon separability (also discernability) of features between no-fault and faulted conditions. The Probability of False Alarm, P(FA), and the Probability of Detection, P(D), are correlated because both are measured with respect to a particular threshold applied to the reduced/fused features. If the threshold is raised to decrease the probability of false alarm, the probability of detection is also inherently decreased. These dependencies are shown in Figure 4.

The Probability Density Function (PDF) of feature values for a no-fault ("normal" or "healthy") condition is on the left, and the range of feature values for a component with a fault is on the right side. The upper left figure emphasizes the no-fault ("healthy") distribution. In this case, P(FA) is on the right side of the threshold, and the Probability of Correct Rejection, P(CR), which represents the range of feature values that (correctly) would not have produced a fault indication given the threshold shown, is to the left. The bottom left plot emphasizes the fault ("unhealthy") distribution. Here, the P(D) can be seen on the right side of the threshold. These are the feature values that would have correctly indicated that a fault existed. The miss rate or Probability of Missed Detection,

FIGURE 4 Statistical feature analysis.

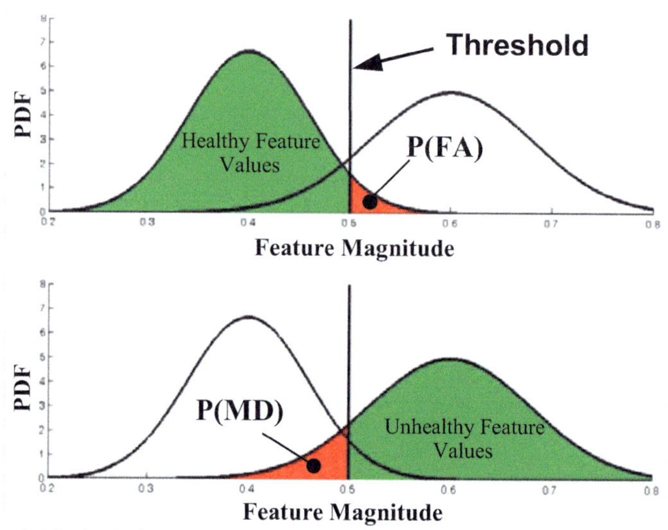

FIGURE 5 CDF representation of feature statistics.

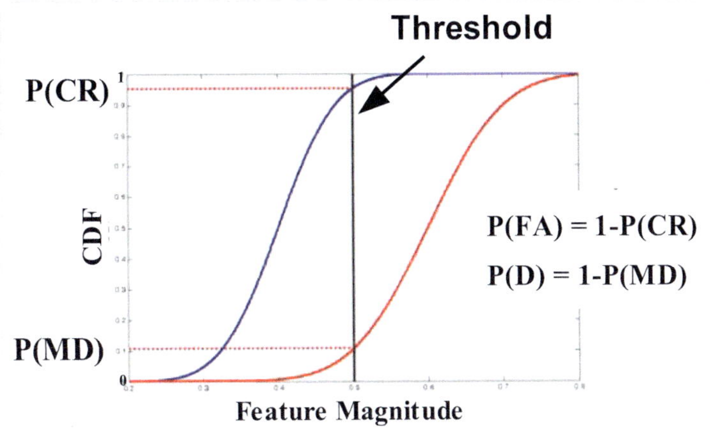

P(MD), is the area below the threshold and represents feature values that would not have indicated a fault given the threshold set point. To decrease P(FA), which is typically desired, we would need to increase the threshold (move it to the right); however, this has the unfortunate effect of decreasing the probability of detection. Figure 5 is another illustration of the inherent tradeoff between P(CR) and P(MD) and also includes the basic relevant equations. In this example, the Cumulative Density Function (CDF) is plotted over the range of features for both the no-fault and faulted cases. Additional information regarding application of these techniques can be found in [12, 13].

GENERATOR VHF HEALTH MANAGEMENT DEVELOPMENT

The authors performed baseline and seeded faults tests using an in-house generator test rig to collect VHF vibration data that could be used to develop techniques for machine fault detection and prediction (Figure 6). The rig consists of an adjustable speed 10 HP drive motor driving a three-phase generator through a drive shaft, which is coupled to the motor shaft through spider couplings. The motor is controlled by a Variable Frequency Drive (VFD).

The generator test stand was instrumented to allow for measurement of VHF vibration. This instrumentation includes high-bandwidth piezoelectric accelerometers and a laser vibrometer, as seen in Table 1.

FIGURE 6 Generator test stand overview.

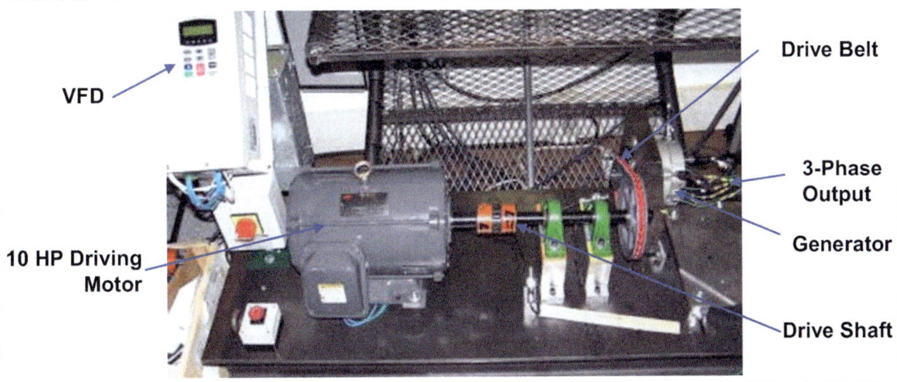

TABLE 1 Generator VHF instrumentation

Sensor type	Bandwidth (kHz)	Location	Sample rate
Accelerometer (PCB 353B16)	70	Front bearing	200 kS/s
Accelerometer (PCB 353B16)	70	Rear bearing	200 kS/s
Laser vibrometer (Polytec)	500	Front bearing	1 MS/s

Three bearing seeded faults were performed on the test stand (each fault was seeded into a different bearing, that is, the front bearing was replaced each time with a different faulted bearing). Fault seeding was complicated by the inability to disassemble and reassemble the bearing; therefore, only faults that did not require disassembly were used. The first fault was seeded by applying a 3/32 in. diamond-tip Dremel bit to the outer raceway surface, producing a number of small dents on the outer raceway. The second fault was seeded into the outer raceway by drilling through the external casing of the bearing using a 1/16 in. carbide drill bit. Finally, a third fault was seeded by inserting 1/10th of a gram medium-grit lapping powder, and manually rotating the bearing at slow speed approximately 100 times to cause the silicon carbide to damage the inner surfaces of the bearing. The bearing was subsequently degreased (to remove the lapping material) and repacked with clean grease. This procedure resulted in minor scuffing of the rotating parts in the bearing. Full investigation of the fault produced would require bearing disassembly (rendering it useless), and was therefore not performed. The three bearing seeded faults can be seen in Figure 7.

Data was analyzed using both conventional signal processing techniques and VHF techniques, including the authors' ImpactEnergy™ bearing PHM algorithm. The results showed a clear advantage of using the ImpactEnergy™ algorithm and amply demonstrated the potential of VHF monitoring to extend the detection horizon of common bearing faults on generator systems. Figure 8 shows sample Fast Fourier Transform (FFT) plots obtained from conventional analysis and the ImpactEnergy™ signal. As the figure shows, the fault frequency of the bearing outer race (BPFO) is not distinguishable above noise for the conventional analysis. However, this frequency is clearly identifiable after applying the ImpactEnergy™ algorithm, indicating that a bearing fault is present.

Furthermore, as seen in Figure 9, the conventional analysis was only able to detect the most severe fault (1/16″ diameter hole, shown in red). The other faults are not distinguishable from the healthy case (note that an increase in this feature is expected). The ImpactEnergy™ results, on the other hand, proved very effective at detecting the bearing faults. In fact, the ImpactEnergy™ feature was clearly separable for the least severe fault (scuffing case, shown in cyan) and was nearly two orders of magnitude higher than the baseline feature for the more severe faults (note that the y-axis of the second subplot is in log scale).

FIGURE 7 Bearing seeded faults, including scuffing (left), 3/32″ dent (middle, internal view), and 1/16″ hole (right, external view).

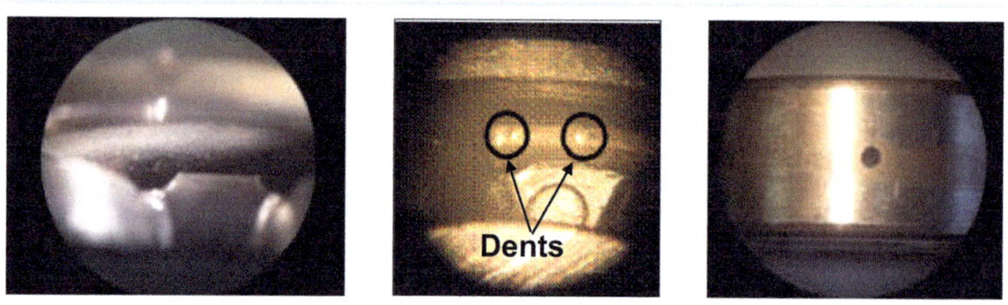

FIGURE 8 Comparison of conventional analysis vs. ImpactEnergy™ FFT plots.

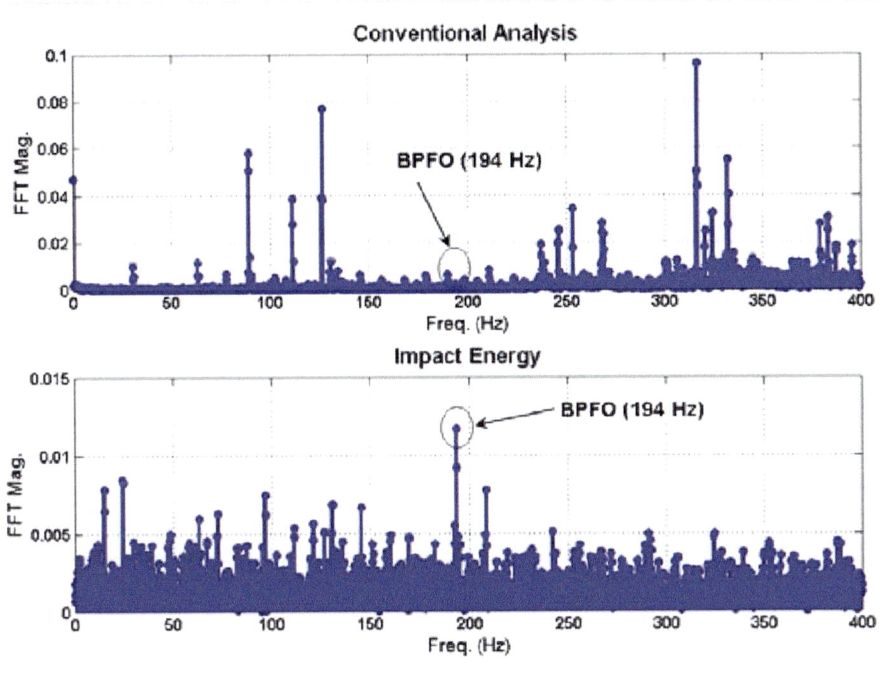

FIGURE 9 ImpactEnergy™ and conventional results from generator bearing fault.

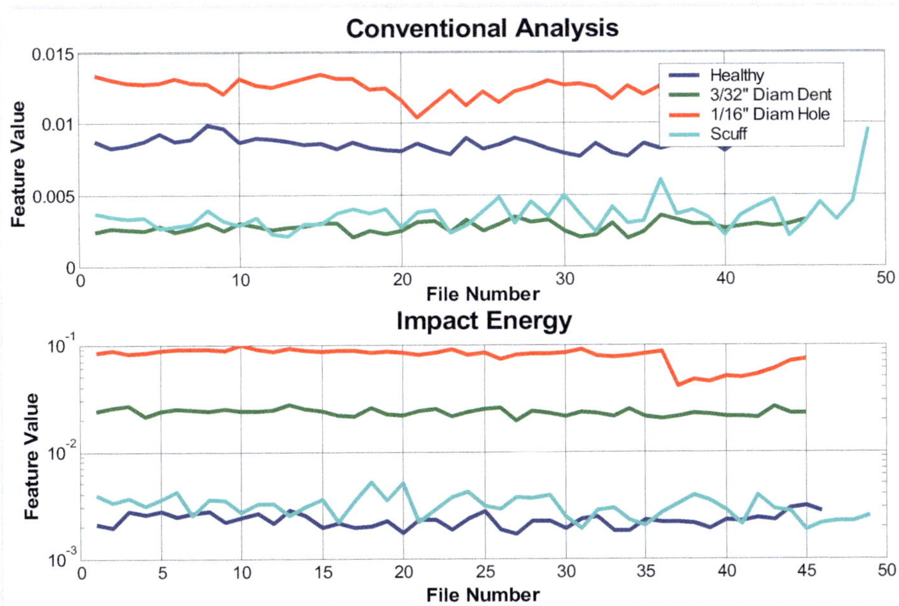

FIGURE 10 Statistical analysis of conventional and ImpactEnergy™ feature results.

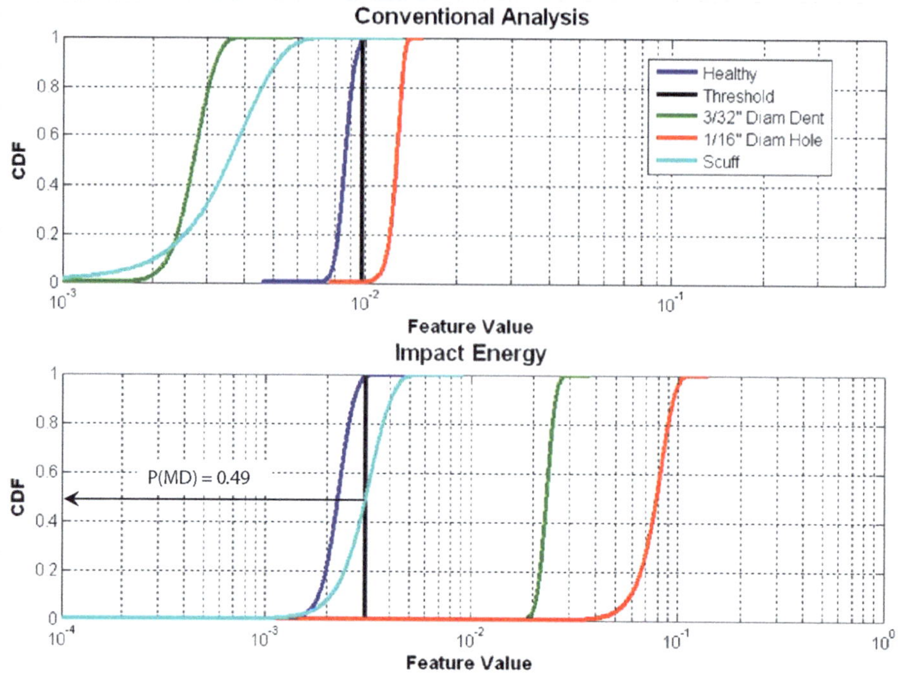

Statistical analysis was also performed and used to compare the performance of the ImpactEnergy™ algorithm against conventional bearing analysis. As seen in Figure 10, the ImpactEnergy™ feature was much more separable for the fault cases analyzed than the conventional feature (note that the x-axis in the figure is a log scale). For this analysis, a P(FA) of 1% was specified and the baseline data (blue curve) was used to determine the threshold (black line) that would be needed to produce this P(FA). Using Figure 10, the P(MD) for each fault can then be determined by evaluating the intersection of the faulted curve with the threshold.

Table 2 shows the P(MD) for these cases. As seen, the conventional analysis performed very poorly for the 3/32″ dent and scuffing faults. ImpactEnergy™, on the other hand, had a relatively low probability of missed detection rate for the scuffing fault (this was a very minor fault), and virtually zero probability of missed detection for the remaining, more severe cases.

ELECTRICAL FAULT SIMULATION

A number of electrical faults were also simulated on the generator test stand. The authors applied their FirstCheck™ algorithm to identify and classify two particular faults,

TABLE 2 Probability of missed detection for conventional and ImpactEnergy™ approaches

	P(MD) for various bearing outer race faults		
Algorithm	**Scuffing**	**3/32″ diameter dent**	**1/16″ diameter hole**
Conventional analysis	1.00	1.00	0.0017
ImpactEnergy™ analysis	0.49	~0	5.30e-010

FIGURE 11 FirstCheck™ feature 1 results for voltage and current signals.

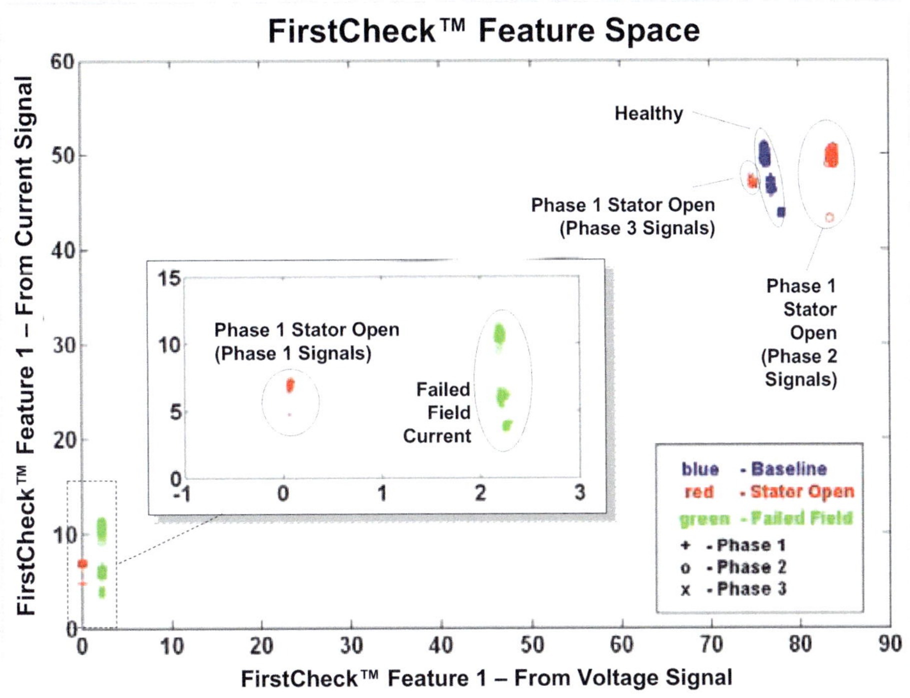

namely, an open stator and failed field current faults. The open stator fault was meant to simulate the fault that would occur if the stator should lose its connection inside the system. It was simulated by disconnecting the positive output from generator phase 1 from the resistive load. The second electrical fault was meant to simulate a failed field current. The test stand rotor is powered by a 12 V, 1 amp power supply that is connected to the alternator through a side terminal and grounded through the alternator housing. The failed field current fault was simulated by unplugging this power supply before operation, thereby preventing the electromagnetic induction of a field current. Although both faults are representative of an actual fault, both represent complete failure and should be readily detectable. As seen in Figures 11 and 12, FirstCheck™ proved very effective at detecting the electrical failures.

In the figures, two FirstCheck™ features are plotted for the voltage and current signals from each phase of the generator. As seen, these features are very separable for the failed field current (green) and the stator open faults (red), which would allow for accurate detection of these faults.

GEARBOX VHF HEALTH MANAGEMENT DEVELOPMENT

The authors also performed baseline and seeded fault tests for VHF algorithm development on an in-house gearbox test rig. The Gearbox Test Stand consists of a 1 HP motor driving a small industrial gearbox (Figure 13). A VFD controller allows motor operation from slow speeds up to 3600 RPM. An intermediate shaft allows additional bearings to be incorporated in the system, increasing the variety of faults that can be studied. The gearbox, which contains a right angle straight bevel 2-gear mesh with a speed reduction ratio of 1.5:1, is driven by a V-belt and is splash lubricated using an aerospace grade

FIGURE 12 FirstCheck™ feature 2 results for voltage and current signals.

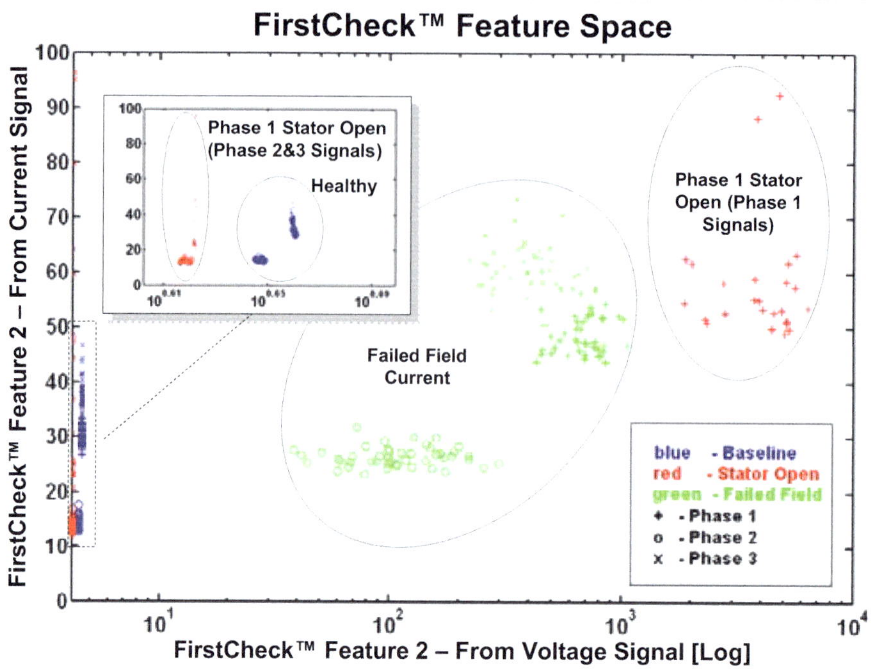

lubricant that meets military specifications. Faults can be simulated in the system using components such as bearings and gears that have seeded defects. Provisions are also included for collecting data with a variety of sensors.

The rig was instrumented with VHF sensors, including two high-bandwidth (PCB 353B16) piezoelectric accelerometers and two AE sensors (one narrowband and one wideband), as seen in Table 3.

Two separate gearboxes were seeded with a corrosion fault to quickly achieve a more severe level of pitting damage. In both cases, gear tooth pitting was chemically induced by subjecting selected teeth on the driving (pinion) gear to a Ferric Chloride ($FeCl_3$) acid solution. Both a mild and a severe fault were seeded, as seen in Figure 14. Baseline and ramp testing were also performed. The faulted gearboxes were then installed on the test rig to evaluate the effect of greater surface degradation levels. Vibration and acoustic data was collected from the two gearboxes while operating at maximum speed and both 15% and 100% torque load. The data was analyzed to determine the effects of the seeded fault on the gearbox vibration signature. In both cases, higher levels of vibration and subsequent increases in various calculated vibration features were observed.

The data collected from the test were used to compare the fault detection capabilities of the various VHF sensors using conventional and

FIGURE 13 Gearbox test stand.

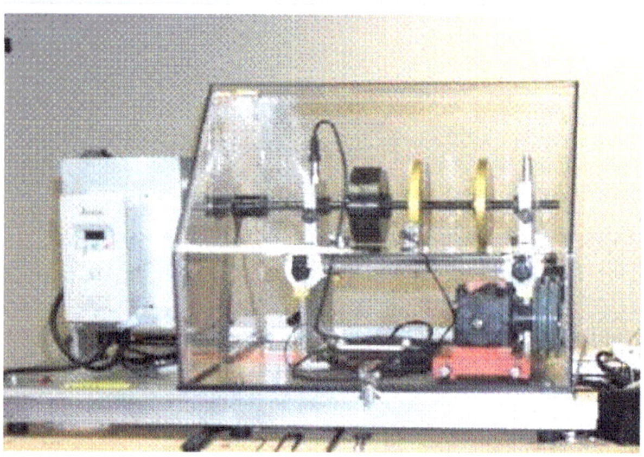

TABLE 3 Gearbox test stand sensor specifications

Sensor	Sensitivity	Freq. range	Resonant freq. (kHz)
Accelerometer (PCB 353B16)	10 mV/g	0.35 Hz-30 kHz	≥70
AE narrowband (Physical Acoustics Corp. R30α)	−62 (dB, ref. 1 V/μBar)	100-400	330
AE broadband (Physical Acoustics Corp. WSα)	−62 (dB, ref. 1 V/μBar)	100-1000	650

innovative VHF approaches. Both vibration and AE data were collected from gearbox baseline and seeded fault tests. This data, which represents over 245 h of testing, was evaluated using the developed VHF algorithms. First, the FirstCheck™ algorithms were applied to check the integrity of the accelerometer and AE signals. Next, mode detection was applied to filter out any transients in the data and avoid the effects of large changes in operational mode (i.e., speed, load, etc.). This was performed since steady-state operation is preferred. Despite this initial filter, small changes in operational mode can still affect features that represent energy level (i.e., RMS), and as a result, normalization was also applied to reduce feature sensitivity to changes in operating conditions such as speed, torque, temperature, etc. Next, the vibration and AE signals were evaluated. GearMod™ was used to extract gear vibration features from the data within the defined operational mode of the system. Processing was performed for both accelerometers for the driving and driven gears. In addition, advanced signal processing techniques were applied to the narrowband and wideband AE data to extract VHF AE features.

FIGURE 14 Seeded corrosion faults (left, mild; right, severe).

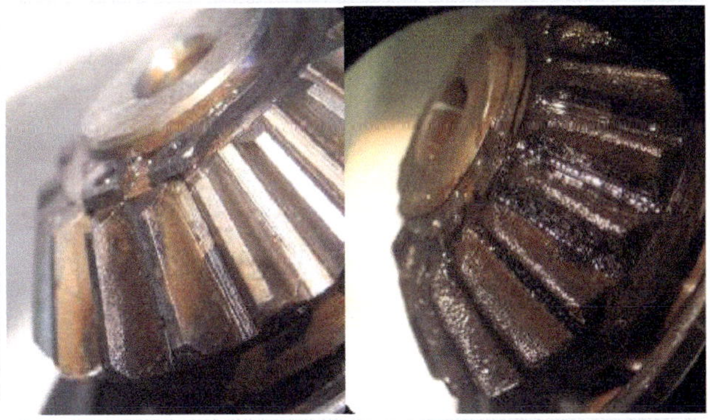

The authors then used baseline and faulty data from healthy and corroded gears to compare the detection capability of vibration and AE approaches. Statistical analysis methods were also used to present the feature PDFs and CDFs, and to calculate the 2% False Alarm Threshold (FAT), P(MD), which is also known as the Missed Detection Rate (MDR), and the P(D), which is also known as the Detection Rate (DR). The upper parts of Figures 15 through 17 are the feature values of healthy (black circles) and corroded (blue circles) data from the front accelerometer, narrowband AE, and wideband AE sensors. The lower left plots in each figure show the PDF representation of the feature in both conditions, including healthy (the black distribution) and corroded (the blue distribution), along with the 2% FAT. The lower right plots show the CDF plots, and include the MDR and DR values.

Figures 16 and 17 show the normalized feature values of the narrowband and wideband AE signals. As seen in the figure, the wideband AE is relatively consistent, while the one of narrowband AE fluctuates and is sensitive to torque changes. These initial results suggest that the wideband AE is more robust and suitable to trend the overall gearbox health condition.

Table 4 summarizes the results of the statistical analysis of each approach. As seen in Table 4, the vibration and narrowband AE approach performed the best, with the

FIGURE 15 Healthy and corroded gear feature response and statistical analysis results with accelerometer data.

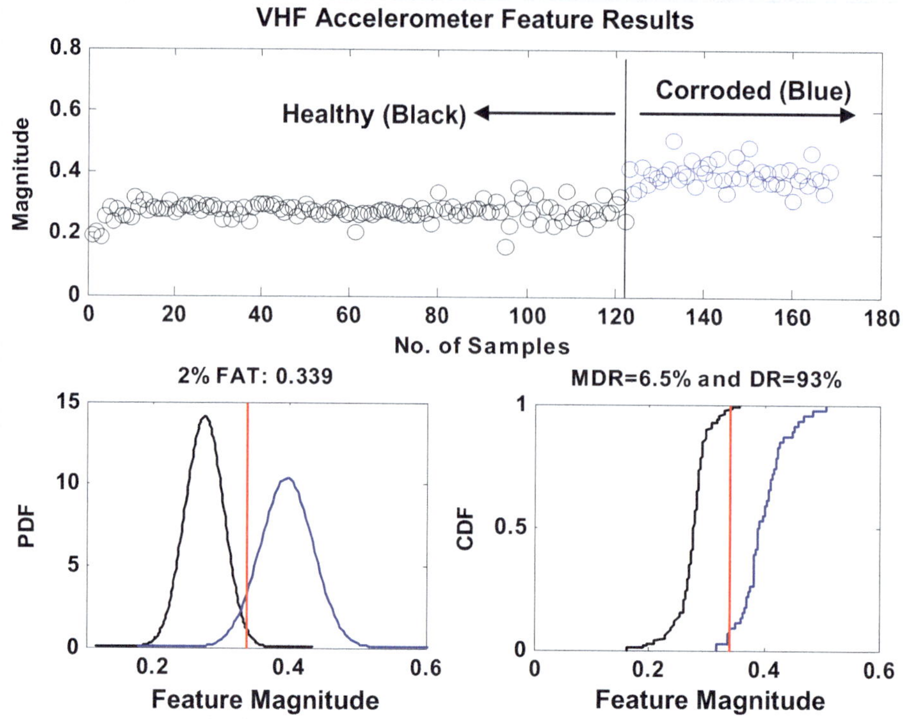

FIGURE 16 Healthy and corroded gear feature response and statistical analysis results with narrowband AE data.

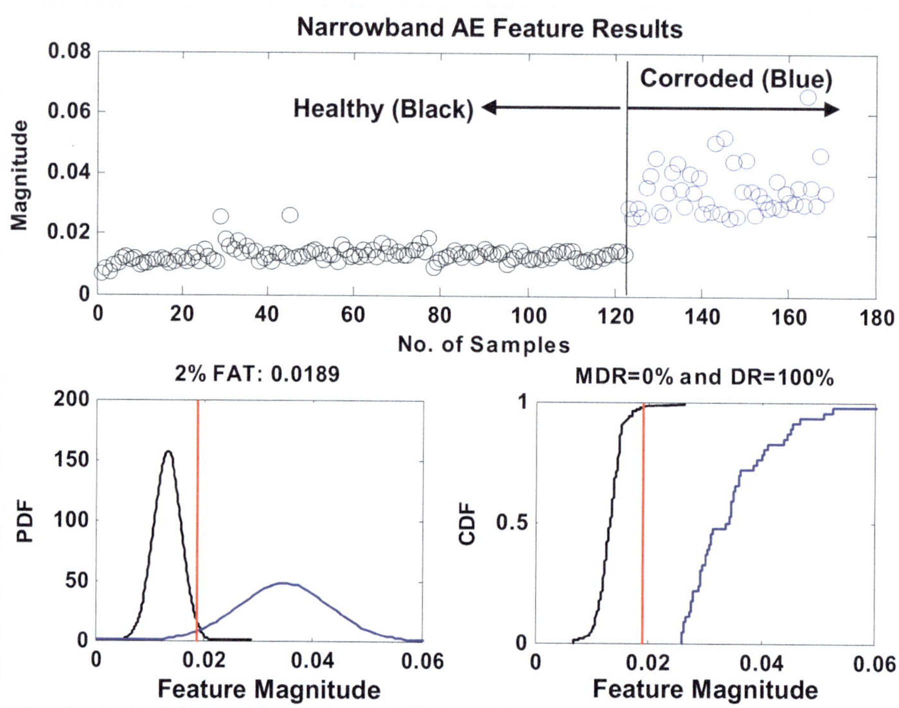

FIGURE 17 Healthy and corroded gear feature response and statistical analysis results with wideband AE data.

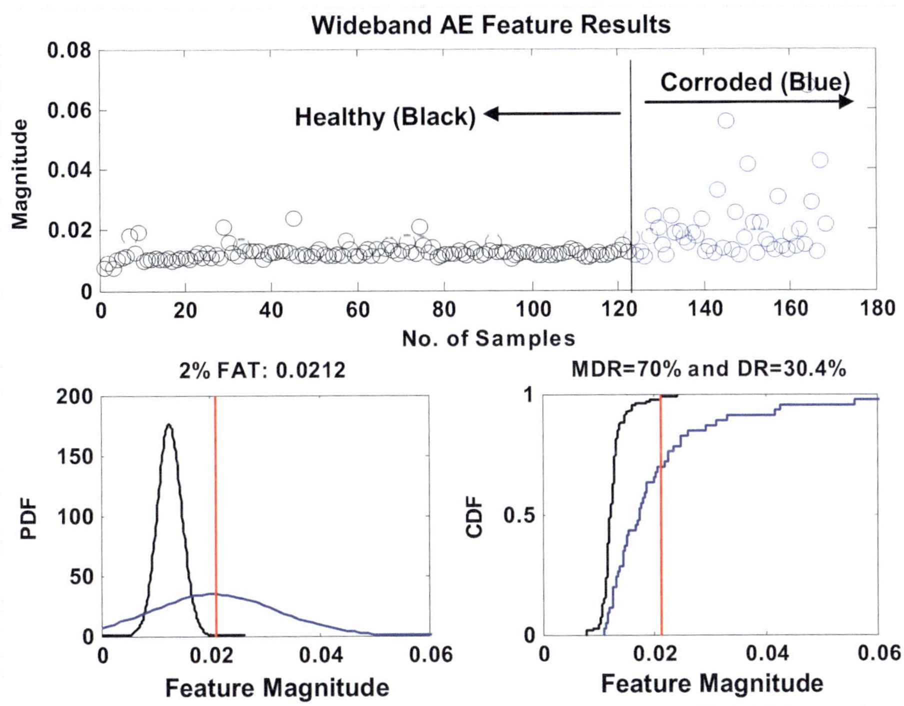

narrowband AE sensor producing the lowest missed detection rate. Although the wideband AE approach showed the worst performance, additional features are being considered to improve its performance. The authors will further evaluate the sensor approaches with other features and select an optimal set of features for more accurate fault detection.

TABLE 4 Gearbox analysis results (with 2% FAT)

Metric	Accel.	Narrowband AE	Wideband AE
Missed detection rate (%)	6.5	0	70
Detection rate (%)	93	100	30.4

Future Work

The authors will continue to mature the developed VHF vibration monitoring and feature extraction algorithms to augment the system's failure prediction capabilities and integrate them with existing data fusion, fault detection logic, and classification algorithms for the automated interpretation of VHF feature sets. Algorithm development will be accomplished through collaborative partnerships with engine OEMs that will provide data collection and implementation opportunities.

The authors are also currently exploring opportunities to test military aircraft engine auxiliary generators in order to develop diagnostic and prognostic (D&P) algorithms for generator health monitoring. The goal of these tests is to collect operational data from military generator(s), including VHF data, in order to develop D&P algorithms to characterize operational signatures of healthy/baseline generators, detect and isolate incipient faults, and prognosticate time to failure based on current fault status.

The authors have also conducted discussions regarding potential collaboration opportunities for accessory gearboxes, including full-scale testing of a military accessory gearbox.

Conclusion

The authors have successfully demonstrated VHF feature extraction using a variety of sensing technologies (piezoelectric, laser vibrometer, and AE). Specifically, the technology has been demonstrated on two subscale test stands. The first is a generator test rig that was equipped with a laser vibrometer and two high-frequency accelerometers. Various mechanical and electrical faults were seeded, with an emphasis on generator bearing faults. Initial results show very good detection capability in frequency bands well above those used in traditional vibration analysis. Another focus, accessory gearbox systems, was addressed for feasibility through a gearbox test rig, which was instrumented with high-bandwidth accelerometers and wideband and narrowband AE sensors. Baseline, seeded fault, and fault progression tests were conducted, including tests with various levels of gear tooth corrosion. Successful detection of this fault was demonstrated using a number of new, innovative approaches. A statistical analysis was also performed to compare the approaches, with narrowband AE and high-frequency vibration features performing the best.

These techniques are being evolved into a unique VHF vibration monitoring system to effectively assess engine gearbox and generator health. The system will be capable of reporting on the early detection and progression of faults for improved incipient anomaly detection. These gas turbine engine vibration monitoring technologies will address existing operation and maintenance goals for current military system and prognostics health management algorithms for advanced engines. These system features will be integrated in a state-of-the-art vibration monitoring system that will not only identify faults more confidently and at an earlier stage but also enable the prediction of the time to failure or a degraded condition worthy of maintenance action.

Acknowledgments

This work has significantly benefited from the technical consult of Ken Semega, Christopher Klenke, and Matthew Wagner of the Wright-Patterson Air Force Research Laboratory (AFRL). The financial support of the U.S. Air Force through a Phase I SBIR contract (FA8650-06-M-2649, "Very High Frequency Vibration Monitoring System for Accessory Health Management") is also gratefully acknowledged. The financial support of previous work, provided through multiple contracts from DARPA and the NAVAIR and Air Force Small Business Innovative Research (SBIR) program offices, is also gratefully acknowledged. This work has significantly benefited from the invaluable support and technical consult of Michael Begin, Andy Hess, Doug Gass, Bill Hardman, and Eric Carney of the Naval Air Warfare Center (NAVAIR) and Joint Strike Fighter program office. Finally, the authors would like to acknowledge the contributions of countless others at Impact Technologies, including Sudarshan Bharadwaj, Hyungdae Lee, Genna Swerdon, Ted Meyer, Chris Minella, Sanket Amin, Jeremy Sheldon, and many others who have contributed to previous efforts and helped make these efforts successful.

References

1. Lebold, M., Campbell, R., Maynard, K. et al., "Review of Vibration Analysis Methods for Gearbox Diagnostics and Prognostics," *54th Meeting of the Society for MFPT*, Virginia Beach, May 1-4, 2000.
2. Bagnoli, S., Capitani, R., and Citti, P., "Comparison of Accelerometer and Acoustic Emission Signals as Diagnostic Tools in Assessing Bearing Damage," *Proceedings of the 2nd International Conference on Condition Monitoring*, London, May 1988, 117-125.
3. Campbell, R.L., Byington, C.S., and Lebold, M.S., "Generation of HUMS Diagnostic Estimates Using Transitional Data," *13th International Congress and Exhibition on Condition Monitoring and Diagnostic Engineering Management*, Houston, December 3-8, 2000.
4. Bagnoli, S., Capitani, R., and Citti, P., "Comparison of Accelerometer and Acoustic Emission Signals as Diagnostic Tools in Assessing Bearing Damage," *Proceedings of the 2nd International Conference on Condition Monitoring*, London, 1988, 117-125.
5. Braun, S. and Datner, B., "Analysis of Roller/Ball Bearing Vibrations," *Journal of Mechanical Design* 101 (1979): 118-125.
6. Kallappa, P., Byington, C., Kalgren, P., and DeChristopher, M., "High Frequency Incipient Fault Detection for Engine Bearing Components," *Proceedings of the ASME Turbo Expo 2005: Power for Land, Sea, and Air*, Paper No. GT2005-068516, Reno, NV, 2005.
7. Kalgren, P., Byington, C., and Kallappa, P., "An Intelligent Ultra High Frequency Vibration Monitoring System for Turbomachinery Bearings," *Proceedings of the 2004 ASME/STLE International Joint Tribology Conference*, Paper No. TRIB2004-64317, Long Beach, CA, 2004.
8. Braun, S.G. and Seth, B.B., "On the Extraction and Filtering of Signals Acquired from Rotating Machines," *Journal of Sound and Vibration* 65, no. 1 (1979): 37-50.
9. Decker, H.J. and Zakrajsek, J.J., "Comparison of Interpolation Methods as Applied to Time Synchronous Averaging," Technical Memorandum, NASA/TM-1999-209086, ARL-TR-1960, Army Research Lab, Cleveland, OH, 1999.
10. Orsagh, R. and Lee, H., "An Enhancement to TSA and Filtering Techniques for Rotating Machinery Monitoring and Diagnostics," *60th Meeting of the Society for MFPT*, Virginia Beach, VA, 2006, 339-349.
11. Orsagh, R., Lee, H., Watson, M., Byington, C. et al., "Application of Health and Usage Monitoring System (HUMS) Technologies to Wind Turbine Drive Trains," *WindPower 2005*, Denver, CO, May 15-18, 2005.
12. Byington, C., Safa-Bakhsh, R., Watson, M., and Kalgren, P., "Metrics Evaluation and Tool Development for Health and Usage Monitoring System Technology," *AHS Annual Forum 59*, Phoenix, AZ, May 6-8, 2003.
13. Roemer, M., Dzakowic, J., Orsagh, R., Byington, C. et al., "Validation and Verification of Prognostic and Health Management Technologies," *Proceedings of the 2005 IEEE Aerospace Conference*, Big Sky, MT, 2005, 3941-3947.

Contact

Matthew Watson is a Manager of Dynamic Systems at Impact Technologies with 8 years' experience in the design, development, and testing of diagnostic and prognostic systems. He has participated in the design of advanced feature development, fault classification, and dynamic systems modeling techniques for a variety of applications, including gas turbine, flight control, power transmission, drive train, electrochemical, fluid, and hydraulic systems. Prior to joining Impact, Matt worked in the Condition-Based Maintenance department of PSU-ARL, where he focused on model-based PHM development of electrochemical and fuel systems. He has co-authored 23 papers related to advanced sensing techniques, signal processing, diagnostics and control, model-based prognostics, data fusion, and machinery health management and is co-author on 2 patents. He has a degree in Mechanical Engineering from The Pennsylvania State University, is a member of the American Society of Mechanical Engineers (ASME), and is the current Vice Chairman of the Machinery Diagnostics and Prognostics Committee within the ASME Tribology Division. Matt can be reached using the information below.

Matthew J. Watson Manager, Dynamic Systems Impact Technologies, LLC
2571 Park Center Blvd, Suite 1 State College, PA 16801 Phone: 814-861-6273 (x102)
Fax: 814-861-6276
www.impact-tek.com Matthew.Watson@impact-tek.com

Alireza Behbahani is the Senior Aerospace Engineer in Controls and Engine Health Management, Structure and Controls Branch, Turbine Engine Division at the Air Force Research Laboratory, Wright-Patterson Air Force Base, Ohio.
(513) 255-5637
Al.behbahani@wpafb.af.mil

Prognosis and Health Monitoring Systems for Aircraft Engines

William Bense
SNECMA

Operational reliability of aircraft engines can be increased if one can detect signs of damages before failure. Prognosis and Health Monitoring (PHM) consists in detecting these signs and giving the probability of equipment anomaly at a defined horizon. This article presents a global overview of PHM systems. Using a system engineering approach, the needs and the equipment to monitor are clearly specified. The dedicated algorithms and embedded/ground systems split are then defined. Embedded system is mainly limited to data acquisition and data reduction. Ground system contains most of the algorithmic part. First, embedded indicators are normalized in order to be compared flight by flight. Then several kinds of approach can be applied to these indicators: among them, trend analysis toward a predefined threshold or anomaly detection. When an anomaly is detected, a classification algorithm identifies failure signature and associated equipment. Lastly, Remaining Useful Life (RUL) is estimated. This algorithmic process requires a crucial learning phase, a topic dealt with in this article. Another key point is to define failure signatures as aircraft engines are extremely reliable. If requested, probability of failure is assessed by an expert. In case of confirmation, customer is notified with the need for maintenance before the estimated RUL.

Introduction

Delays and cancelations are both expensive for aircraft companies and "irritant" for passengers. Prognosis and Health Monitoring (PHM) consists in detecting signs of damages before failure and giving the probability of equipment anomaly at a defined horizon. The aim is to predict failures before they happen in order to give the engine operator the ability to plan the maintenance action. Operational events can then be reduced. SNECMA builds a set of technologies to deal with premises of abnormalities. R&T team is working on bench test data, operational aircraft data retrieved from the flight data records, and simulation of every engine systems. The research work led to a set of algorithmic solutions using several interesting mathematic tools. The algorithms partly described below and mostly referenced in past articles generate diagnosis indications as risk of damage, failure probability, and result precision. All those outputs are linked together in a final decision layer which generates notification reports. This whole process is implemented partly on an embedded computer and partly on a ground monitoring system. Mathematical techniques used in Health Monitoring area may be reviewed, for example, in referenced papers [3, 6, 11] or [12].

This document aims at giving the engineer point of view and explains how to use the mathematical algorithms as proficient tools for making diagnostics and prognostics on engine aircraft health. The article begins by detailing the system engineering approach used. Then it describes the embedded and the ground systems main function by main function from data acquisition to advisory generation. Lastly, it details how to measure and manage algorithm quality.

System Design

A system engineering approach has been followed to define the needs, the perimeter of the PHM system, and its functionalities.

<u>The first question of all system approaches is always "what is the need?"</u>

- The need is to be able to plan maintenance actions and then increase operational reliability of the aircraft engines and reduce maintenance costs.
- Another aspect is to detect dormant failures which do not affect flight security but increase wear of other equipment or fuel consumption.
- Safety aspects are out of PHM scope and are managed differently according to their own certification rules.

<u>Which equipment to monitor?</u>

- Failure Mode and Effect Analysis (FMEA) studies and lessons learned highlight most probable failures and associated equipment. Their consequences in terms of In-Flight Shutdown (IFSD), Delays and Cancellations (D&C), fuel consumption, etc. shall be considered.
- Frequent maintenance actions like oil filling are also an important aspect of PHM. Indeed, optimizing these frequent maintenance operations can save money.

<u>Aims of equipment health monitoring</u>

FMEA highlights potential failures. Then the link between failure modes and data observation should be established.

- The state of the art for each equipment monitoring shall be considered. Every already existing company's technology should be taken into account.
- Mathematical and physical principles of their surveillance must be defined. This is based on observations and physical simulation. Such studies lead to the identification of exhaustive sets of health indicators.

<u>How to monitor these equipment</u>

- Define required sensors. Is there a need for new sensors addition or is it possible to use existing sensors with eventual algorithm adaptation?
- Define embedded/ground split strategy: Embedded hardware and software are more expensive, and the process to develop and update them is stronger than for ground systems.
- Define data transmission and hardware capability compatible with embedded/ground split strategy.

All these points are not addressed linearly. The process is iterative and a compromise between all these items must be found: for example, reducing embedded hardware and software triggers an increase of air-ground data transmission capabilities. Functional analysis is a useful tool for our system engineering approach and a help to manage our system's complexity.

System engineering has led us to define precisely the system needs. Now, let us describe the resulting system, first with its embedded part and then in its ground one.

Note: To describe the PHM system and functions, the OSA-CBM Standard is commonly used (see reference [1] for its complete description) which recommends splitting our system in the following steps:

1. Data Acquisition
2. Data Manipulation
3. State Detection
4. Health Assessment
5. Prognosis Assessment
6. Advisory generation

Each layer of OSA-CBM decomposition will be addressed in this article with explanations and corresponding examples. Only layer 2 which includes the first data preprocessing steps spans over embedded and ground systems.

Embedded System

As previously stated, the embedded system is reduced to a minimum because it is more expensive and more difficult to update than the ground system. But there are parts that cannot be exported into the ground system:

- Data Acquisition which is intrinsically embedded
- Data reduction (part of Data Manipulation-OSA-CBM layer 2) for an easier air/ground transmission (Reference article [2] also approaches this split aspect)

Data Acquisition (OSA-CBM Layer 1)

Data Acquisition of each requested sensor is performed continuously during the entire mission, but due to the huge amount of data, embedded monitoring algorithms are not executed all the time but only during relevant flight phases.

An Operating Modes detection algorithm is key to recognize accurately flight phases. This flight decomposition is used not only to trigger algorithms but also for user interpretation of data as the flight profile can give useful context information.

Data Reduction (OSA-CBM Layer 2)

Data reduction has two main steps:

- Selection of relevant data for each algorithm. This is done firstly by the flight phase detector and secondly by the algorithm itself.
- Creation of an indicator using these data.

DATA SELECTION AND TRIGGER

Each embedded algorithm is controlled by a trigger. The flight phase detector runs all the time, whereas other algorithms run only when their data and the processor are available.

For each algorithm, a list defines which data are required and during which flight phase.

Let us consider a monitoring algorithm as an example. This example algorithm needs results from 10 sensors among the plurality of data and during Cruise phase only. It also requires stability conditions according to the engine speed for instance. When the flight phase detector algorithm identifies a Cruise, a trigger launches data temporary storage of those 10 sensors in a buffer. Outside Cruise, data is not stored.

As soon as the processor is available, these data are computed by the embedded monitoring algorithm (it is in real time if possible, actually with delay due to the number of algorithms but always before the end of the flight).

The monitoring algorithm receives data from the 10 sensors during the entire Cruise phase. As only data acquired during engine speed stability are relevant, the algorithm selects itself a time interval from the Cruise condition which satisfies the engine speed stability requirement. Cruise data which does not comply with this requirement can be dumped to minimize data buffer size in the embedded equipment.

It must be noticed that this is just an example: there are algorithms working on other flight phases, and some of them are focused on unstable intervals instead of stable.

INDICATORS CREATION

The algorithm computes data when some conditions are satisfied and generates embedded reduced indicators. Example indicators can be average and standard deviation of a temperature during the considered time interval, the extremum of a value, as any sort of computation. The same process is applied for each algorithm: preselection using flight phase information, selection by the algorithm itself and indicators computation. Figure 1 summarizes this process.

At the end of the flight, each embedded indicator is transmitted to a ground station whereas raw data are kept on board and downloaded upon request. At this point there is a plurality of embedded indicators. The ground system will process them and underline abnormalities and trends to failure.

FIGURE 1 Embedded PHM system. Algorithms are triggered according to flight phase. Once launched they select data and build indicators.

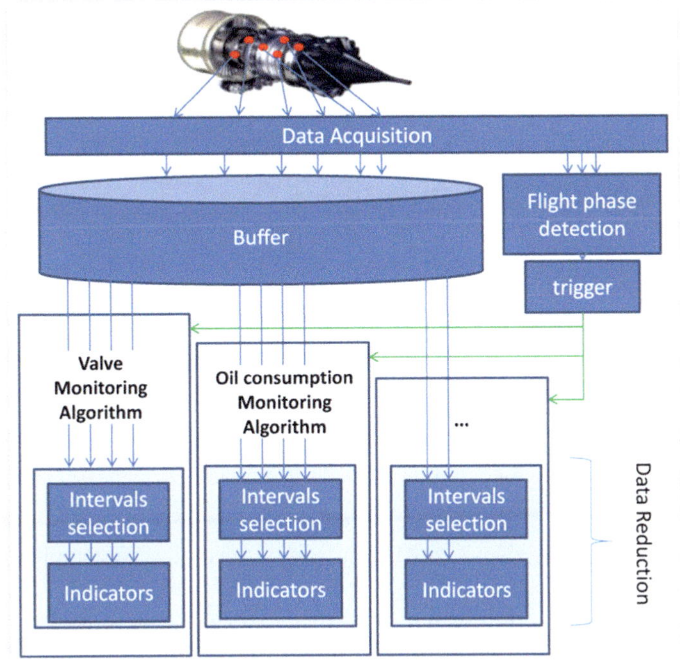

Ground System

Data (previous health indicators) are received and stored in a database. Then they are analyzed by the ground system.

Normalization (OSA-CBM Layer 2)

Even if embedded indicators have been acquired during similar conditions (engine speed stability algorithm for example), they cannot be compared as these acquisition conditions are not strictly the same (e.g., stability at 90% of max rpm is different from stability at 85%).

Normalization is required to extract this context influence (flight dependency). Depending on the algorithm it may be:

- A context normalization (monovariate—only one endogenous measurement at a time): data are normalized with exogenous data (context data like altitude, ambient temperature, etc.) which influence the measure.
- A full normalization (multivariate—simultaneous endogenous measurements): data are normalized using exogenous but also other endogenous data (i.e., indicators of system state and not only context information).

Figure 2 illustrates these differences. Usually a context normalization simplifies the physical interpretation while a full normalization takes into account dependencies between measurements but is less interpretable.

Reference article [3] deals with oil temperature normalization and explains how to choose between endogenous and exogenous data.

FIGURE 2 Example of the difference between monovariate and multivariate normalization.

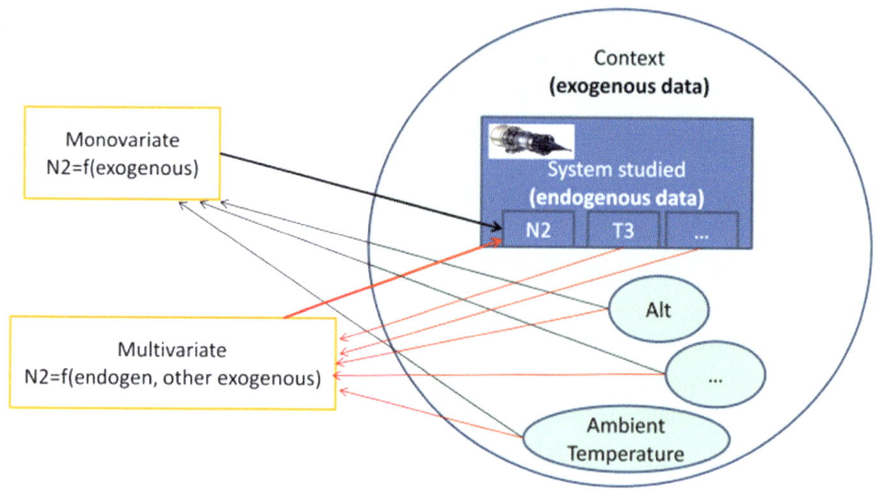

Reference article [4] gives another example of normalization, applied to Exhaust Gas Temperature (EGT).

Note that normalization usually requires a learning phase to determine links between data (except if a physical model is used).

Choice of reference flights used for learning is really important. Indeed, flights used for this phase must not have abnormalities; hopefully aircraft engines are so well maintained that we have lots of healthy observations and definitely enough for normalization purpose.

Usually the first flights may be chosen for training update as a new engine rarely has failures. During the engine life, new learning updates may be useful to adapt to new use conditions (an engine state may change even if there is no failure).

Launching a new learning phase must also be considered when the physical engine configuration change. Two initial learning methods are possible, and a choice should be made between personalized or generalized learning:

Engine specialized learning means the N first flights are used as reference: the advantage is that normalization is customized and links between variables are not theoretical.

→ Normalization is much better in that case. The problem is that the learning is limited to the behavior cases met during these first flights; outside these cases, normalization quality is not good.

Generalized learning phase: The other possibility is engine type learning, in which the entire fleet is used as reference.

→ This method for normalization has exactly the dual advantages/disadvantages to the specialized solution. Choice depends on the application.

Figure 3 summarizes how to choose between generalized and customized normalizations.

Indicators Analysis (OSA-CBM Layer 2)

Normalized data can be compared and followed flight after flight. Depending on the kind of equipment and associated degradation, two approaches may be considered: anomaly detection or trend analysis.

FIGURE 3 Choice between an engine customized normalization and a generalized fleet normalization.

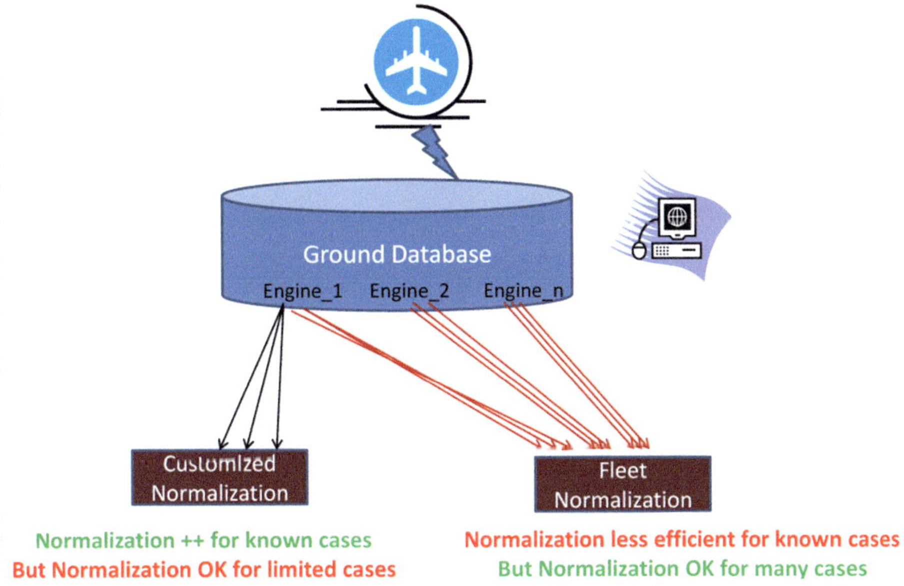

ANOMALY DETECTION

First of all, a model of normality must be learned. This learning phase defines a reference value and standard deviation (problems of the learning phase are exactly the same as for normalization).

For each new value, the difference between the current value and the learned reference is estimated (as illustrated in Figure 4, a score is calculated to estimate the ratio between the measured deviation and the standard deviation).

Each indicator is associated with an individual score.

But a Health Monitoring algorithm monitors a plurality of indicators and it is necessary to quantify the global deviation. This can be made using a Mahalanobis distance. This distance is equivalent to Euclidean distance when data are independent and takes into account correlations between data.

Mahalanobis distance is given by:

$$d^2 = (y - \mu)' \sum^{-1} (y - \mu)$$

where

d is the Mahalanobis distance
y is the reference value
μ is the current value
Σ is the covariant matrix

For more details about Mahalanobis distance, consult reference documents [5]. Reference article [3] gives an example of global score using Mahalanobis distance and its evolution cycle by cycle to detect engine abnormality.

TREND ANALYSIS

Trend analysis consists of monitoring the time evolution of normalized indicators by estimation of a slope.

FIGURE 4 Building a score.

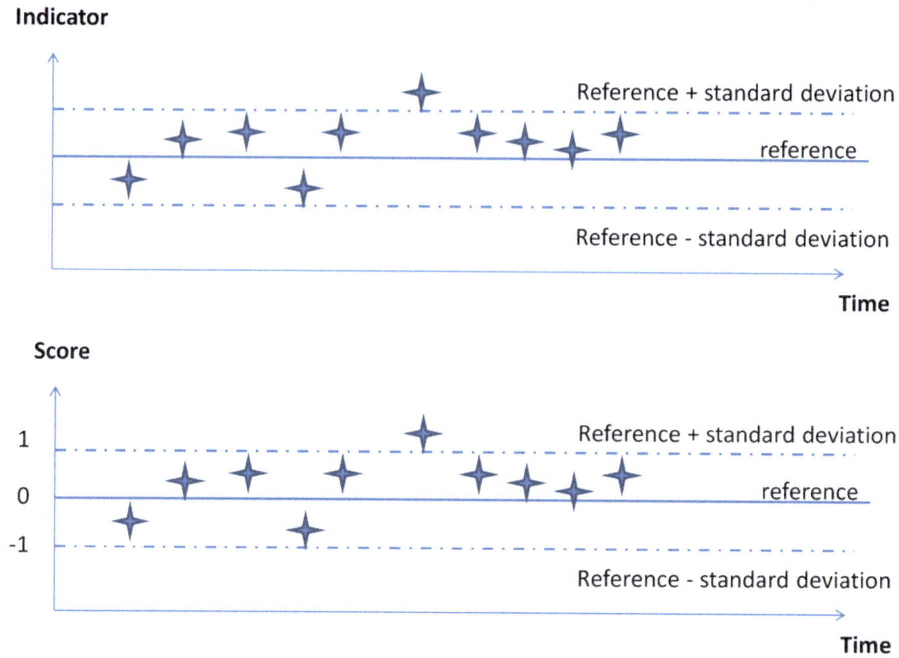

If the behavior of the indicator is not linear, it can be transformed with a model: for instance, we will monitor a logarithm for an exponential behavior.

A shift in this slope will be revealed in the State Detection layer (next step of OSA-CBM): a change in behavior indicating a sign of failure as illustrated in Figure 5.

State Detection (OSA-CBM Layer 3)

The Data Manipulation layer built scores or slope indicators.

In the State Detection layer, these scores/indicators are compared to predefined thresholds.

FIGURE 5 Trend sudden change can reveal an abnormality.

For example, for a score, a deviation higher than 3 times the standard deviation can be chosen to be a sign of failure (for a Gaussian law, 99.7% of values belongs to the interval [-3σ; $+3\sigma$]) as presented in Figure 6.

Reference article [6] deals with detection of sudden changes and how to detect them among the noise. At this step abnormalities are detected but are not linked to any particular equipment.

CLUSTER ANALYSIS

Another way to detect abnormality is to use Self-Organizing Maps (SOM).

SOM are a type of neural network where each node corresponds to a neuron locally organized in

FIGURE 6 Threshold use on indicators permits to detect abnormalities.

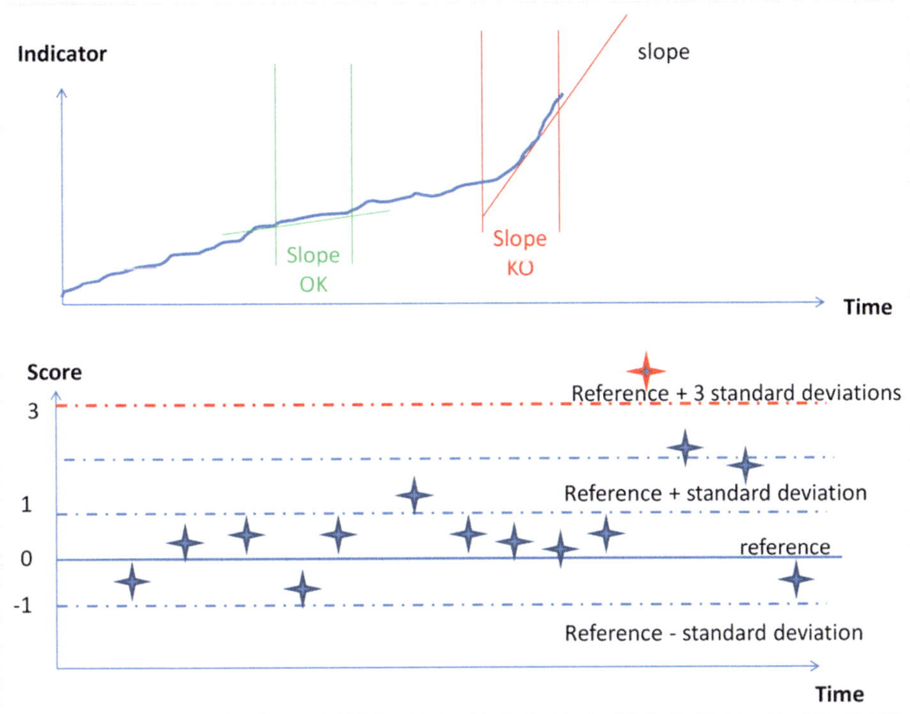

a 2D map. Each neuron will identify the center of a cluster of similar observation in the original high-dimensional space of features.

First a map is created and the position of each node is randomly distributed. The algorithm takes the first point of the learning phase and searches the closest node, increase the weight of this node, and increase a little the weight of closer nodes. This algorithm makes this process iteratively and obtains a map.

After the learning phase, it projects the engine onto this map flight by flight and detects if it has an abnormal trajectory. Figure 7 gives an example of sudden trajectory change.

SOM can also help to identify classes of engines using experience. If an engine has the similar trajectory of a past engine with problems, this engine has a high probability to encounter similar problems.

Reference article [7] details how to detect abnormality in SOM with an example where EGT change behavior is superior to other indicator changes.

FIGURE 7 Trajectory of an engine on a SOM. Inside the green circle the beginning of the trajectory and in the red one a sudden change revealing abnormality.

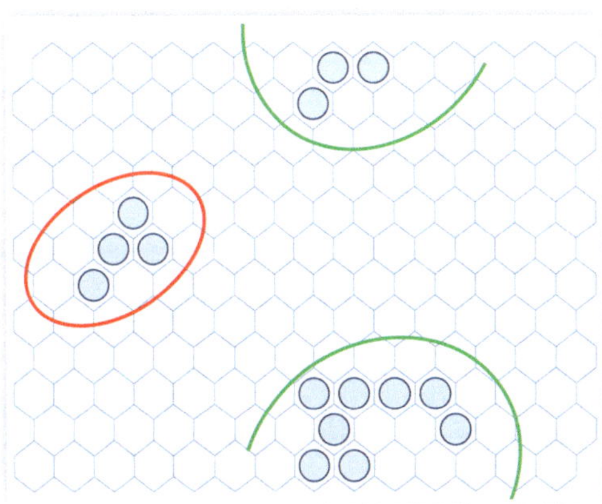

Health Assessment (OSA-CBM Layer 4)

Once an abnormality is detected, the next step is to identify which equipment is damaged and what is its state.

Some indicators are explicit: for example, "valve opening duration" anomaly directly identifies the valve. But some indicators are less explicit and an abnormality can be observed on a plurality of data. A failure classification algorithm is used.

FIGURE 8 Experts define caricature of failure signatures. Failure can be indicated by a limited number of parameters (as Caricature 3, with indicator 14).

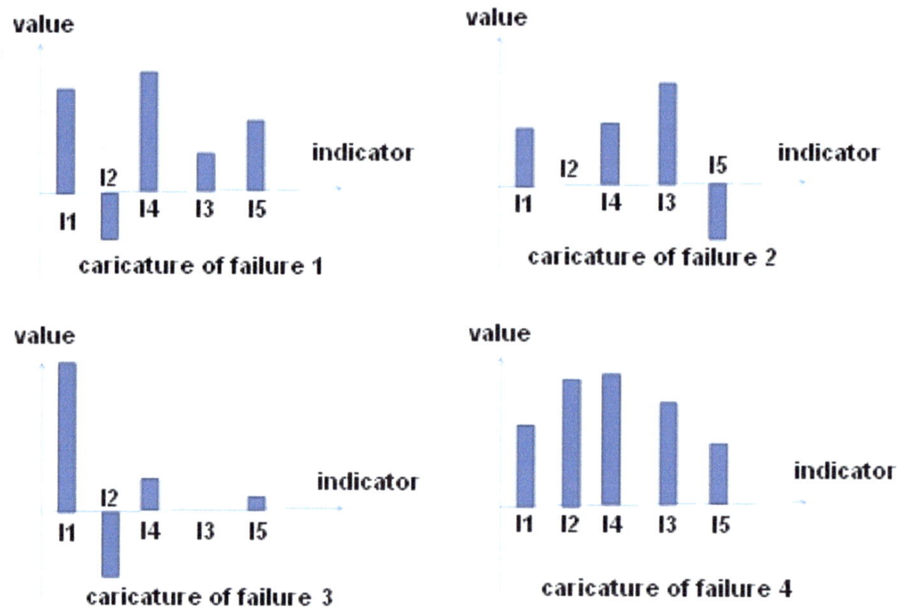

Assuming there is an abnormality, this algorithm identifies failure signatures. First, a matrix of failure signatures is built up based upon using past experience and experts' knowledge. This matrix can use a lot of indicators.

Figure 8 gives an example of four failure caricatures defined by five indicators.

When an abnormality is detected, its signature is projected in the n-dimensions sphere. An approximation of the geodesic distance between each signature and the caricatures is estimated as illustrated in Figure 9, using the cosine of the angle between the caricature and the signature. It permits to estimate the probability of the current signature to be one of the reference failure signatures.

FIGURE 9 A 3D representation of classification algorithm and associated distance.

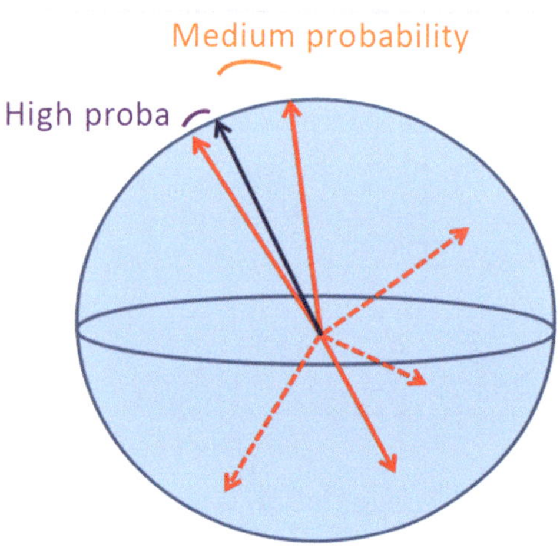

The algorithm gives the probability of each failure signature. Now, it is necessary to identify which system is abnormal as the abnormality can be observed on several signatures.

This step is based on Bayes' rules and conditional probabilities: during a learning phase, experts define the conditional probability of equipment failure if signature s_i happens. With this information, the algorithm can finally give:

- The probability the current signature represents reference signature ref_i
- The probability the current signature represents a failure of equipment eq_i

Reference article [3] applied classification to engine start sequence monitoring and proposes graphical view of its results.

Note that a general problem for Health Monitoring algorithms maturation and for learning phases is to have data with failures: engine aircraft are extremely reliable and failures are very rare. Real failure signatures are difficult to obtain, that is why expert knowledge to define failure caricatures is required. It is not dealt with in this article, but indicators fusion can also be used to strengthen diagnosis (see reference article [10]).

Prognosis Assessment (OSA-CBM Layer 5)

Once signs of failures are identified and the associated equipment determined, it is necessary to estimate the RUL.

The way to estimate this duration really depends on the failing system and the specificity of failure. A physical aspect is very important but mathematics can help to approximate this remaining time. In Figure 10 a trend extrapolation is used to estimate RUL.

Another possibility of trend extrapolation use is to determine the probability to cross a threshold at a predefined horizon. Indeed, customers often need to know if they can use the aircraft in the next *N* flights.

FIGURE 10 Trend extrapolation algorithm can help to determine RUL.

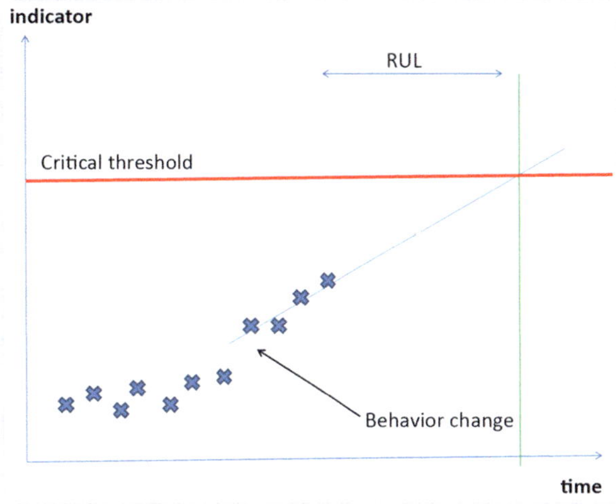

Examples given in Figures 11 and 12 use linear extrapolation. Considering a normal distribution around the critical threshold, the extrapolated trend at the defined horizon gives the probability to cross this threshold. If the probability is higher than a predefined value, an alert is generated. If the behavior of the system is known, the problem can often be reduced to a linear one: for example, if the behavior evolution is exponential, log (indicator) is linear.

If the behavior is unknown, a sampling algorithm like particle filter (also called Sequential Monte Carlo method) is used after identification of temporal behavior of the system.

It is important to notice for this part, but also for all the previous steps of this article, that proposed algorithms deal with probability: a sign of failure is determined as a probability to fail higher than a predefined level at a given horizon.

FIGURE 12 Trend extrapolation: in this example there is a high probability to cross the threshold at the predefined horizon.

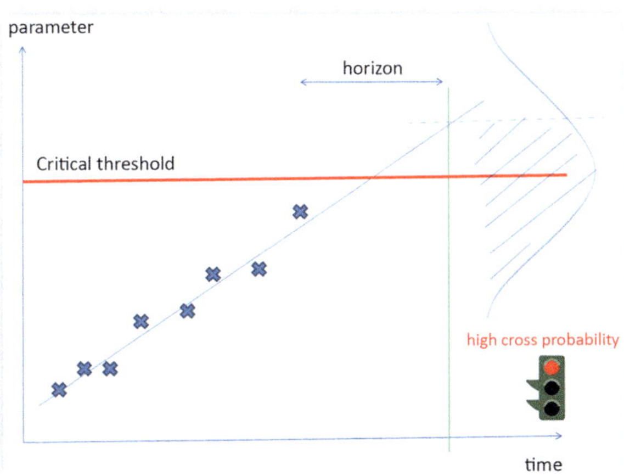

FIGURE 11 Trend extrapolation: in this example there is little probability to cross the threshold at the predefined horizon.

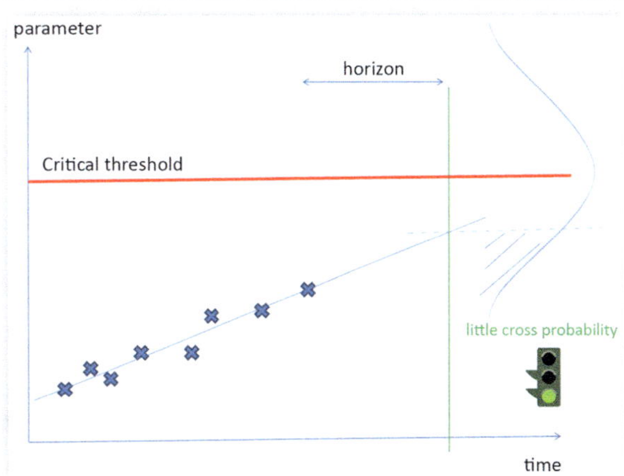

CHAPTER 10

FIGURE 13 General process of PHM system—embedded and ground.

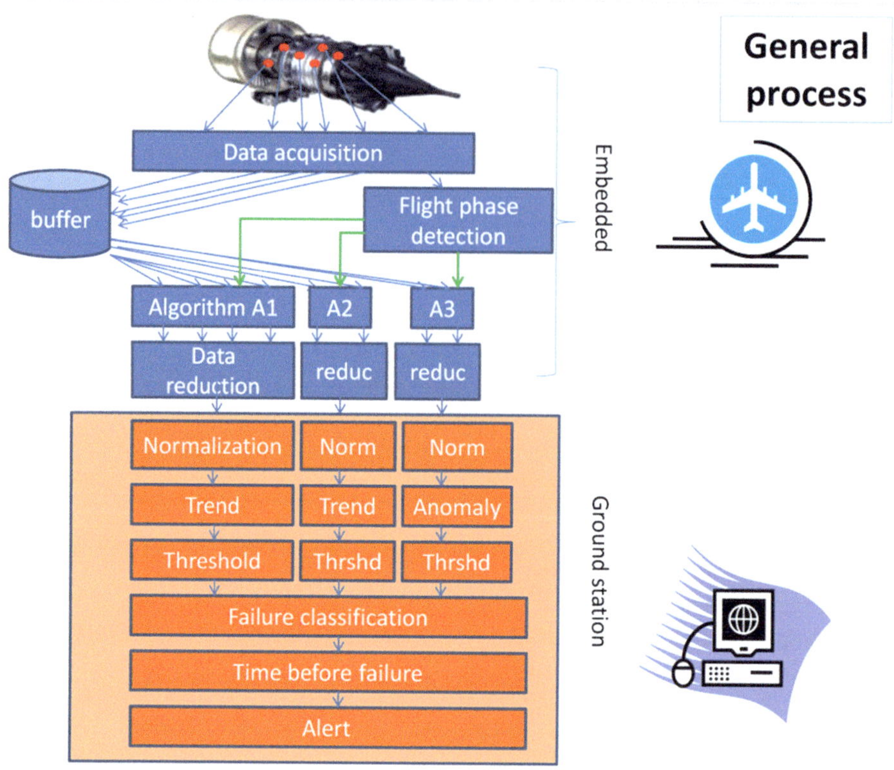

Advisory Generation (OSA-CBM Layer 6)

All the previous algorithm information can generate an alert to inform the Customer Support Center (CSC) about the probability of failure/need for maintenance action of all the monitored equipment. As algorithms deal with probabilities and statistics, nothing is certain and an alert is never sent directly to the engine owner. Each alert and the associated RUL are confirmed or disconfirmed by the CSC. In case of confirmed sign of failure, a Customer Notification Report (CNR) is sent to the engine owner which has the ability to plan maintenance actions. Figure 13 summarizes the general process from embedded data acquisition to alert generation.

Many of the previous mathematical techniques are based on probability: they do not give a 100% confidence. The following paragraph details how to measure their quality and then how to adjust the algorithms and associated alert thresholds.

Algorithm Quality

Each class of engine is different from others and each algorithm must be adapted. Several Key Performance Indicators (KPI) have been defined to measure their quality. The two main KPI are:

- Probability of Detection (POD)
- Probability of False Alarm (PFA)

They give information about the quality of the PHM algorithms. Indeed a system giving too many false alarms will be rejected by the user, but a system with bad rate detection is useless too.

Let us consider the example of trend extrapolation (Figures 11 and 12). There is an alarm if there is a high probability to cross the threshold, but what is the limit between a high and a low cross probability? Adjusting this limit will have a significant effect on PFA and POD: if the CSC is alerted when there is a probability higher than 80% to cross the threshold, PFA will be higher and POD smaller than if this limit is set to 90%.

Indeed, with 80% threshold, the algorithm will generate more alerts, but the confidence in their correctness will be smaller than with a 90% threshold. That is why adjusting alert thresholds precisely is really important.

A large number of other KPI has been developed—as Adequation Quality Value (AQV) which quantifies the robustness by computing the adequation between the current flight(s) and flight(s) used for learning phase. Reference article [8] can be consulted to have a more complete list of KPI. Reference article [9] details the use of KPI applied to engine oil consumption monitoring algorithm.

Conclusions

PHM system is a key service to optimize maintenance actions and reduce operational events. But this kind of system is complex and split into embedded and ground parts. Managing and developing these PHM systems relies on two important aspects: the first is system engineering to organize data flow and algorithm layers and the second is the mathematical aspect to develop the tools required to increase prognosis and diagnostic efficiency. This article highlighted both of these key aspects and the associated difficulties.

References

1. MIMOSA, "Open Systems Architecture for Condition-Based Maintenance," OSA-CBMv3.1 Standard, 1998.
2. Serge, B. et al., "Health Monitoring des moteurs d'avion," *Les entretiens de Toulouse 2009*, 2009.
3. Jérôme, L., "Standardized Failure Signature for a Turbofan Engine," *IEEE Aerospace Conference*, Big Sky, 2009.
4. Schimert, J. and Wineland, A., "Coupling a Dynamic Linear Model with Random Forest Regression to Estimate Engine Wear," *Annual Conference of the Prognostics and Health Management Society*, 2010.
5. Prasanta, M., "On the Generalized Distance in Statistics," *Proceedings of the National Institute of Sciences of India*, 1936.
6. Jerome, L. and Etienne, C., "Sudden Change Detection in Turbofan Engine Behavior," *CM & MFPT*, Cardiff, 2011.
7. Etienne, C., Marie, C., Michel, V. et al., "Aircraft Engine Health Monitoring Using Self-Organizing Maps," *Industrial Conference on Data Mining*, New York, 2010.
8. Jerome, L., "Validation Environment of Engine Health Monitoring Algorithms," *IEEE Aerospace Conference*, Big Sky, 2012.
9. Jean-Rémi, M., Ouadie, H., and Xavier, B., "System PHM Algorithm Maturation," *First European Conference of the Prognostics and Health Management Society*, 2012.

10. Jerome, L., "Emergent Decision Reports Using a Fusion Protocol," 2013.
11. Jérôme, L. and Nya, D.R., "Detection of anomalies in an aircraft engine," Patent WO 2010061136, 2010.
12. Jérôme, L., "Identification of defects in an aircraft engine," Patent WO 2010076469, 2010.
13. David, C., "Condition Monitoring of Gas Turbine Engines," Technical Report, Department of Engineering Science, University of Oxford, 2006.
14. Upadhyaya, R. et al., "Multivariate Statistical Signal Processing Technique for Fault Detection and Diagnosis," *ISA Transactions* 2, no. 4 (1990): 79-95.
15. Christopher, B., *Pattern Recognition and Machine Learning*, (Springer, 2006).
16. Yann, P., "La démarche de l'ingénierie système," jfod.cnam.

Definitions/Abbreviations

CNR - Customer Notification Report
CSC - Customer Support Center
EGT - Exhaust Gas Temperature
IFSD - In-Flight Shutdown
FMEA - Failure Mode and Effect Analysis
KPI - Key Performance Indicator
MRO - Maintenance, Repair, and Operations
OSA-CBM - Open Systems Architecture Condition Based Maintenance
PHM - Prognosis and Health Monitoring
rpm - rotation per minute
RUL - Remaining Useful Life

CHAPTER 11

Aircraft Integration Challenges and Opportunities for Distributed Intelligent Control, Power, Thermal Management, and Diagnostic and Prognostic Systems

Alireza R. Behbahani and Alex Von Moll
Air Force Research Lab.

Robert Zeller and James Ordo
Rolls-Royce Corp.

Modern propulsion system designers face challenges that require that aircraft and engine manufacturers improve performance as well as reduce the life-cycle cost (LCC). These improvements will require a more efficient, more reliable, and more advanced propulsion system. The concept of smart components is built around actively controlling the engine and the aircraft to operate optimally. Usage of smart components intelligently increases efficiency and system safety throughout the flight envelope, all while meeting environmental challenges. This approach requires an integration and optimization, both at the local level and the system level, to reduce cost. Interactions between the various subsystems must be understood through the use of modeling and simulation. This is accomplished by starting with individual subsystem models and combining them into a complete system model. Hierarchical, decentralized control reduces cost and risk by enabling integration and modularity. This process involves defining, developing, and validating against requirements for key integrated propulsion, power, and thermal management system capabilities.

Introduction

Continuous onboard situational awareness of the vehicle state using prognostic health management (PHM) technology will enable integrated vehicle control. It will also improve the safety and reliability of the aircraft by performing self-diagnosis and self-correction in flight. The development of adaptive turbine engine technologies will provide optimized fuel efficiency and increased performance capabilities. Controlling the engine more efficiently requires smart, distributed, optimized controls and PHM. Optimized control of the propulsion system includes the electrical power generation by the engine and thermal management systems (TMS), integration to the vehicle control and health management system, and adaptive features that adjust to the mission and equipment health status.

The integration of propulsion control systems and flight control systems has shown significantly improved aircraft performance parameters. When the propulsion, power, and thermal subsystems are not integrated, each subsystem must be able to operate in a worst-case combination with other subsystems, requiring large operating margins [1]. In this paper, it is shown how large operating margins contribute to the thermal management issue. To maximize the performance and efficiency of the total aircraft, each system functions and cooperates with one another in a precise and timely fashion, as shown in Figure 1.

Current practice in the aircraft engine industry is to locate the full authority digital engine controller (FADEC) physically close to or on the engine (i.e., On-Engine concept). To the engine manufacturers, this means that the product supplied to their customer can be thoroughly tested before delivery, with a warranty covering items that are under total control of the engine suppliers. Responsibility for any engine or related control system failure clearly rests with the engine supplier. For new decentralized FADEC components, as well as aircraft components, adaptive engines required to improve specific fuel consumption are assumed to be improved separately. The decentralized smart components still require a centralized master controller for the initial partially distributed system, where some higher level control functions are contained in a processing unit physically remote from engine (i.e., Off-Engine concept), and collocated with flight

FIGURE 1 Integration of complex systems in order to accomplish the desired mission.

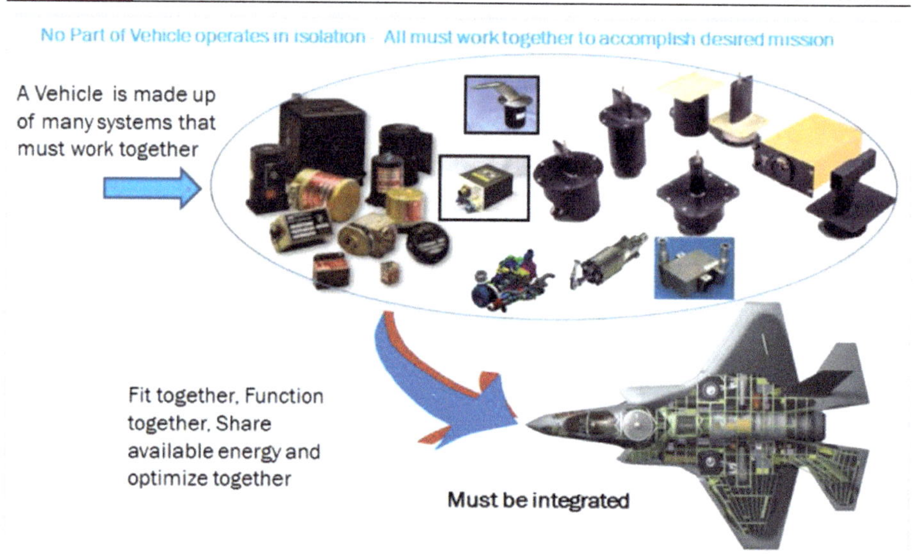

control and possibly the mission avionics processing unit. The Off-Engine design will have the additional communication buses between the processors and the engine bus terminal. Redundancy and routing of these buses should be designed to have no significant negative effect on survivability. Whereas the On-Engine approach does tend to simplify the contractual obligations of the many parties involved in the design, integration, testing, and operation of aircraft engines, it does not lead to reduced LCC benefits for the customer. Due to the complexity of design and manufacturing process for engines, second sourcing may be prohibitively expensive. However, the development cost of engine control for the decentralized distributed controls for the Off-Engine approach can be drastically reduced in the long run. The Off-Engine concept has the potential to improve and advance the controls technologies for all aircraft as well as engine manufacturers, while standardizing the design. A number of enabling technology advancements which include electronics miniaturization, addition of local intelligence to sensors, increasing high-temperature electric actuator capabilities, increasing processing capability, and increased integration between the engine(s) and the flight control technologies are being investigated that should further improve the safety performance and efficiency of the aircraft as well as the comfort of passengers. The improvements that can be achieved through innovative software and algorithms and the hierarchical distributed controls are the focus of this paper. It concentrates on areas that do not require significant advances in hardware (e.g., high-temperature components including processors, sensors, and actuators) to make the improvements achievable. Even greater benefits can be achieved when those technologies become available. The evolution of aircraft has focused on the need to enhance basic capabilities, including range, payload, speed, and operating cost, all of which have been partially served by improvements in control systems, prognostics, and diagnostic technologies. The operational need of new and existing aircraft and missions has made more demand on propulsion controls.

As a result, propulsion control systems must manage multiple effectors based on multiple operating parameters through interactive processes. Modern propulsion control systems are increasingly integrated with TMS, power, PHM, and aircraft flight controls, and the distinction is becoming less and less noticeable. Distributed control of the future engine will enable greater integration of propulsion control with aircraft flight control and subsystems: thrust reverser, variable geometry inlet and exhaust, pressurization and environmental control, power, hydraulics, and thermal management, especially for the adaptive engines. Distributed, decentralized control system integration defeats the centralized approach of limiting and managing control system complexity. Conventional federated control system design approaches become difficult and costly to define, integrate, validate, and sustain. Through hierarchical control systems, in which a set of devices and governing software are arranged in a hierarchical tree, as well as adaptive controls, one can incorporate each component separately and independently while seamlessly integrating each subsystem.

Challenges for Management Systems and Integration

Electrical Power Challenges

Next-Generation aircraft missions continue to require additional electrical power to power airframe systems and mission payloads. This trend has accelerated as directed energy and higher power communication suites are deployed into service. Optimization

FIGURE 2 Electrical power challenges [3].

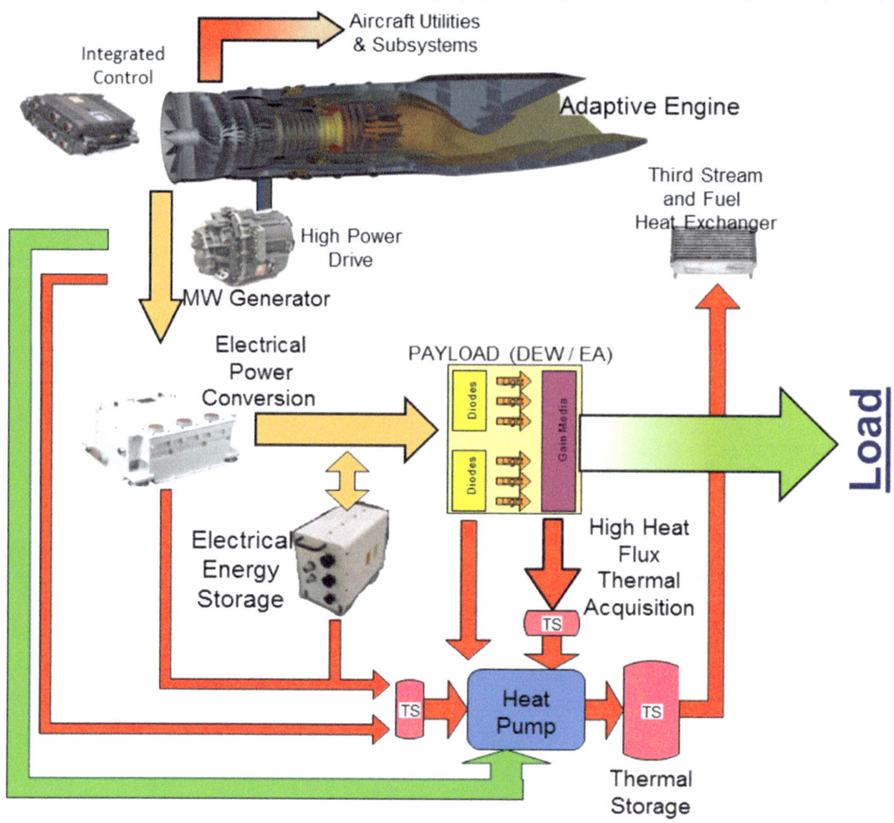

of future systems will require complete understanding of the power requirements across the aircraft as the maximum power offtake increases and load profiles vary throughout the mission. Future power system architectures have demonstrated the benefits of platform wide integration trade studies which integrate the traditional power flows—propulsion, electrical, and thermal—to optimize system performance. Study results indicate [2] that future aircraft designs will require reconfigurable or adaptive systems to increase the overall energy efficiency (range), reduce the number of components (size and weight), and improve the reliability at the platform level. Because these electrical loads are large and dynamic, new electrical storage subsystems are being developed to reduce the transient impact on the stability of the electrical system. The physical integration of these new systems allows increased power density but results in increasing control complexity (Figure 2).

Thermal Management Challenges

As new platforms have been developed, the Air Force and its contractors have focused insufficient attention to system-level thermal management. Thermal loads are increasing while available heat sinks are decreasing and the thermal challenges have become increasingly difficult to manage. The ad hoc approach to thermal management has resulted in operational limits on fielded platform systems such as the F-22 and F-35B, and if not properly addressed, will create limits on future long-range strike, directed energy weapons, and intelligence, surveillance, and reconnaissance systems (Figure 3).

FIGURE 3 Thermal management challenges [3].

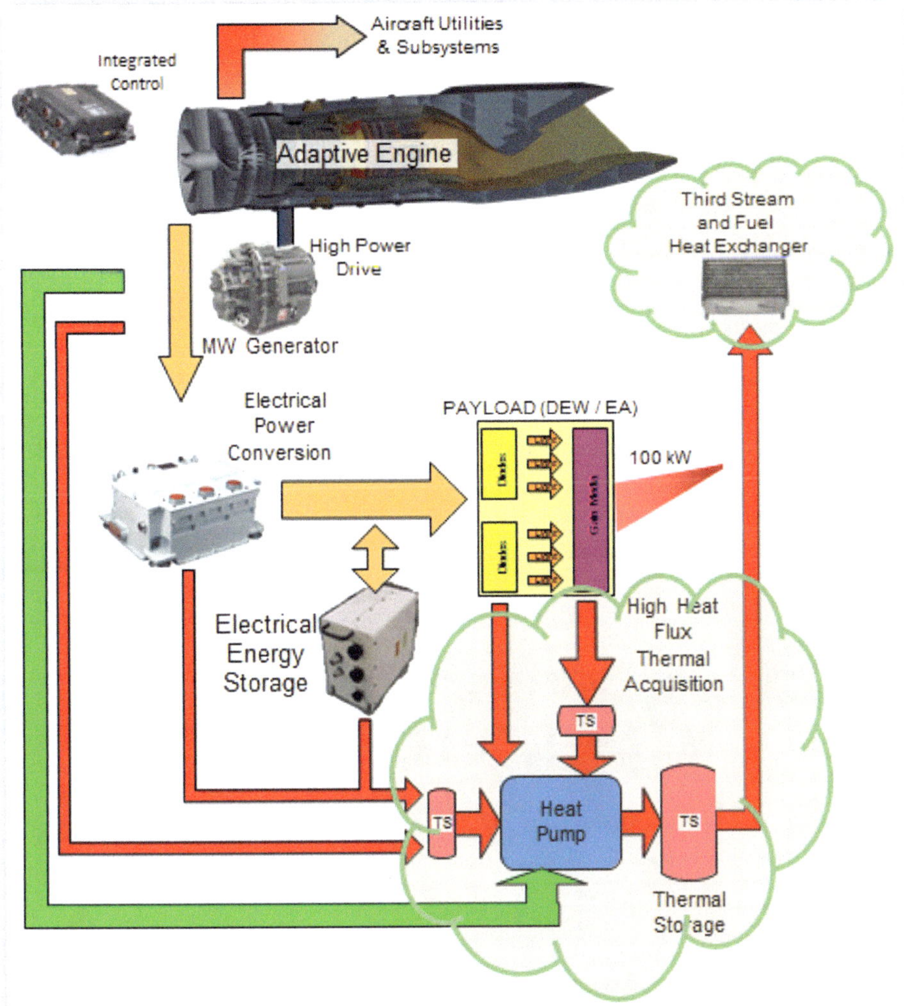

Engine Control and PHM Challenges

The future challenges for turbine engine controls are (1) sensors implementation for turbine engines that run at higher temperature and (2) better accuracy for better performance and specific technologies for diagnostics, stability management, and reconfiguration for damage tolerance. These challenges include tip clearance control, active combustion control, data fusion, integration of the turbine engine with the flight control, and high-frequency analysis for turbine engine controls. These technologies will prolong the life of the engine, increase reliability and maintainability goals, and reduce LCC. In all techniques, the objective is to reduce engine wear and tear and obtain maximum life from the components. The Distributed Engine Controls Working Group (DECWG) has identified the technical challenges for the distributed engine control to be [4]:

1. Communication pathway from FADEC to engine system: bandwidth requirements and how is it affected by the engine environment
2. Power distribution to sensors and actuators, previously performed by FADEC: how it will affected by the engine environment

3. System controllability-functionality distributed from FADEC to local subsystems: system affected by network latency software partitioning, etc.
4. System modularity: system engineering, integration, verification, and certification paradigm changes

From the user's perspective, additional challenges exist, including ever-increasing requirements for performance, reliability, and cost of ownership. From a design perspective, the primary challenge to the advancement of distributed intelligent engines is the harsh operating environment present in and around the gas turbine engine, which is being addressed by the DECWG.

Integration Challenges

Propulsion aircraft integration presents unique challenges to the development of a propulsion system (Figure 4). Many of these challenges result from the fact that the design of components and subcomponents are made independently of each other, and the

FIGURE 4 Propulsion and flight control systems integration [3].

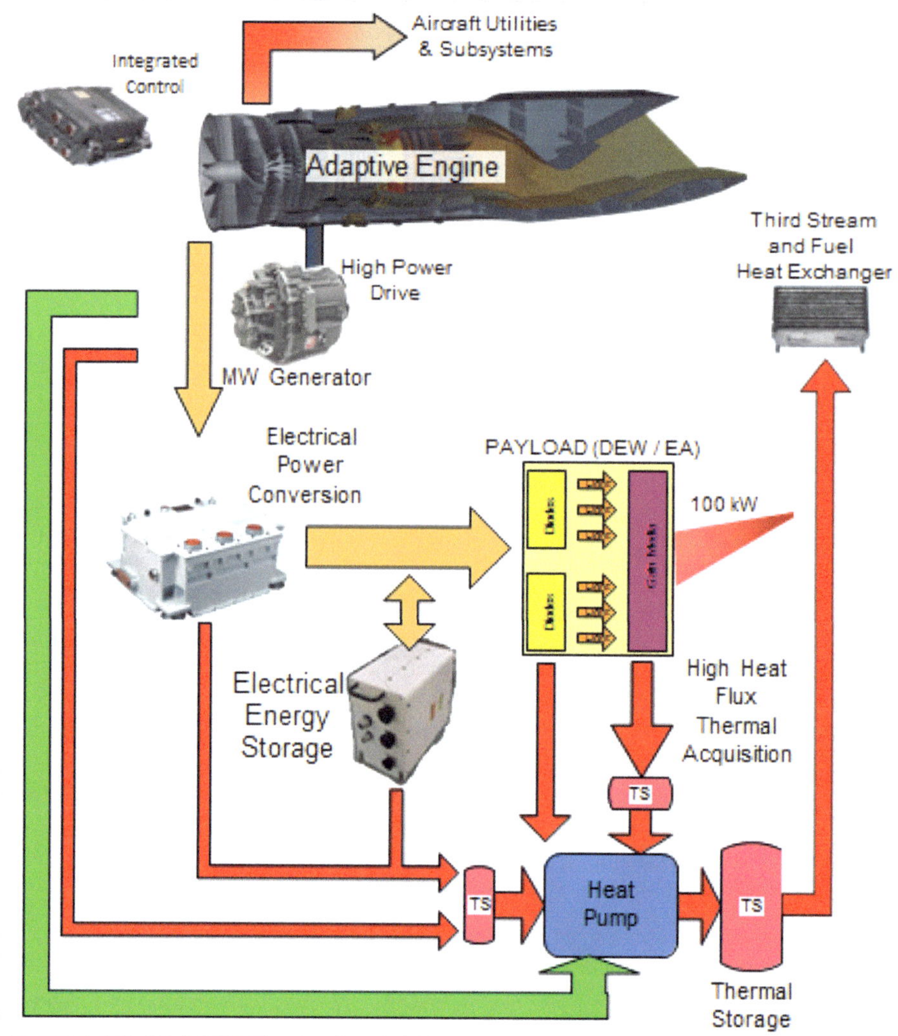

integration issues involve major interfaces between aircraft components and engine manufacturers [5]. High-quality design tools and standardized processes, standard data-bus communication, hierarchical software, and an open system approach will be needed to apply these tools. In the future, the challenge of analyzing advanced designs will dictate the use of advanced techniques for quantitative analysis.

The Versatile Affordable Advanced Turbine Engines (VAATE) Program is a joint DoD/NASA/DOE/Industry effort to develop revolutionary and innovative technologies by the 2017 timeframe and beyond that will permit an order of magnitude increase in turbo-propulsion affordable capability over the year 2000 state-of-the-art technology. The VAATE program mission is to develop, demonstrate, and transition advanced multi-use technologies that provide a revolutionary improvement in affordable capability to a broad range of legacy, pipeline, and future military propulsion and power needs, with explicit versatility for dual-use application. VAATE is focused on total propulsion system capability. The airframers have been actively studying the development and integration of key future propulsion system technologies for air vehicle systems in the Air Force Research Laboratory (AFRL) mission capability areas. The overall objectives of this program are to establish an updated and clear, traceable need for VAATE III and beyond turbine engine technology development and integration plans. These objectives will be met by relying on the large set of engine technology expertise within airframers and the partners who will be involved in these studies, as well as on the vision systems in the key VAATE capability areas that are being leveraged from other sources.

Challenges of integrating a propulsion system on an aircraft, new technologies, and new business drivers dictate innovative approaches be developed and applied to the design process addressing the cost, weight, and size challenges. Technologies will be broken into six major categories: controls, PHM, inlet, exhaust, engine cycle integration, and IPTMS technologies. Estimated technology readiness level (TRL), risk, and priority will be established for each technology with input from all stakeholders, including the end users. Focus will be on air vehicle propulsion integration technologies, with less emphasis on those technologies that exist independent of the vehicle (i.e., cooled-cooling air, turbine blade materials, etc.). Of course, control is the overarching technology that must be included in all aspects of the integration, including many challenges in propulsion system performance, operability, and meeting system requirements.

To meet the challenges ahead for the installation aerodynamics technology discipline, key enabling technologies are needed. Controls, transient analysis, and high fidelity analytical tools, primarily CFD, product data management, and product configuration control are essential.

Total Real-Time Vehicle Integration, Optimization, and Control

Through adaptive control, the control system can be made resistant to variation in components, systems, subsystems, engine behavior, engine/aircraft variation, and component aging. Additionally, one can optimize aircraft performance through a variety of techniques:

- Manually or automatically trimming control constants
- Altering schedules in conventional algorithmic control software
- Synthesizing vehicle state, real time, from measured aircraft, engine, and component characteristics
- Using plant models (in a model-based control)

FIGURE 5 Interconnection of controls for integrated propulsion, power and thermal for next-generation capability.

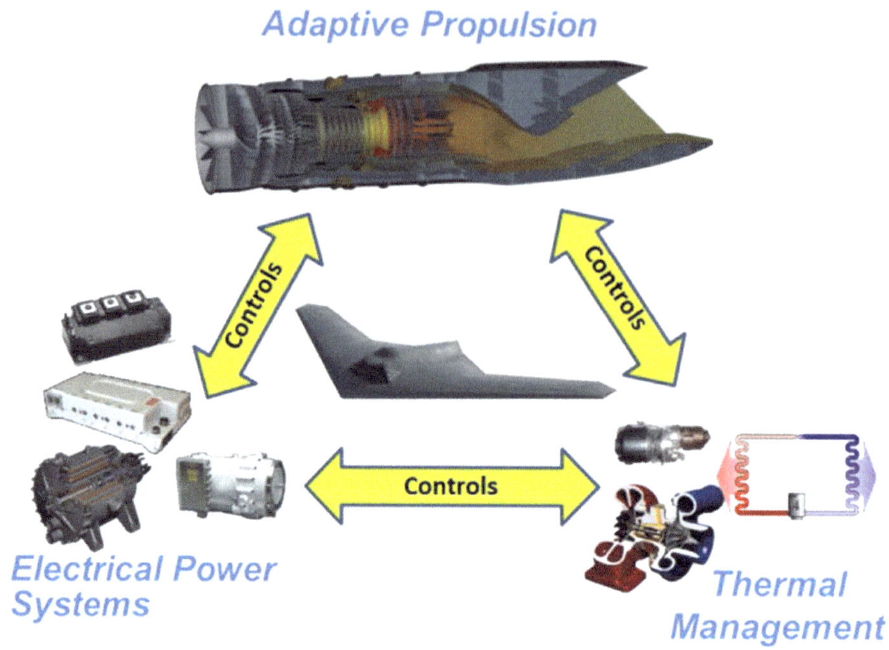

Figure 5 shows the importance of the controls' role for an integrated propulsion, power, and thermal management system to enable next-generation weapon systems. Electronic engine control is the key feature of the improved propulsion control system, engine operability, and effective integration with other airplane systems. Whether its design and/or selection of a single component or full integration of every aspect of an integrated vehicle system, we need to focus on controls in developing and deploying the most advanced power, conversion, distribution, thermal management, and propulsion control technology. The integration of multiple electronic as well as control systems technologies, TMS, and power at systems' level enables significant improvements in the functionality and efficiency of the aircraft system. This system must account for operability, controls, horsepower extraction, thermal loads, electrical loads, vehicle impacts, and mission impacts. In this arrangement, control is the glue that ties all of the different subsystems and components together, allowing everything to function efficiently both as an individual subsystem and in concert with the rest of the aircraft. Past efforts at achieving this level of integration have fallen short in the area of real-time adjustment. Control laws for many different components and subsystems have traditionally been scheduled, a priori, for a predetermined set of flight conditions. Adjusting the interaction between power, thermal, and propulsion subsystems in response to a variety of conditions and measurements in real time is necessary for maximizing the benefit of total vehicle integration.

Hierarchy and Structure

Just as there is stratification in a functional sense, there is a sort of vertical, or hierarchical, layering in these systems as well (Figure 6).

At the lowest level, this could be exemplified by a single smart sensor. This sensor would likely consist of a sensing element, analog-to-digital conversion electronics, and a means of communicating over a data bus. From a control system's perspective, this sensor is providing some sort of measurement of a physical quantity (temperature, pressure, vibration, etc.). From a power perspective, this smart sensor may be consuming some amount of power and is connected with the rest of the electronic grid. From a PHM perspective, this sensor might have a processor built in that is continually checking and reporting the health and status of the sensor itself. Lastly, from a thermal management perspective, it may be that the sensor electronics require fuel cooling and contributes some heat to the rest of the fuel cooling loop. Each of these views of a single smart sensor is important in defining how the smart sensor is designed and incorporated into overall aircraft systems. It should be clear that controls, PHM, power, and thermal management are intertwined into nearly every component and cannot be considered as separate systems which can be optimized independent of one another. This is even truer in scenarios where there are large transients, such as in modern military fighter aircraft. Here, components spend a lot more time operating at their peak conditions. Components fail, temperatures and pressures can change quickly, and quick maneuvers can impose large g-forces—thus, a system that is optimized for a specific selection of conditions is likely to spend most of its time in suboptimal operation. As all of these factors are affecting the aircraft, perhaps more heat will be produced than anticipated, or more power will be needed than anticipated, or a portion of the system has failed and the control needs to enter a reversionary mode of operation. The efficiency and performance of the aircraft will suffer if the control system does not adapt and optimize the components and their interaction in a real-time fashion.

FIGURE 6 Functional and hierarchical relationships in aircraft management systems.

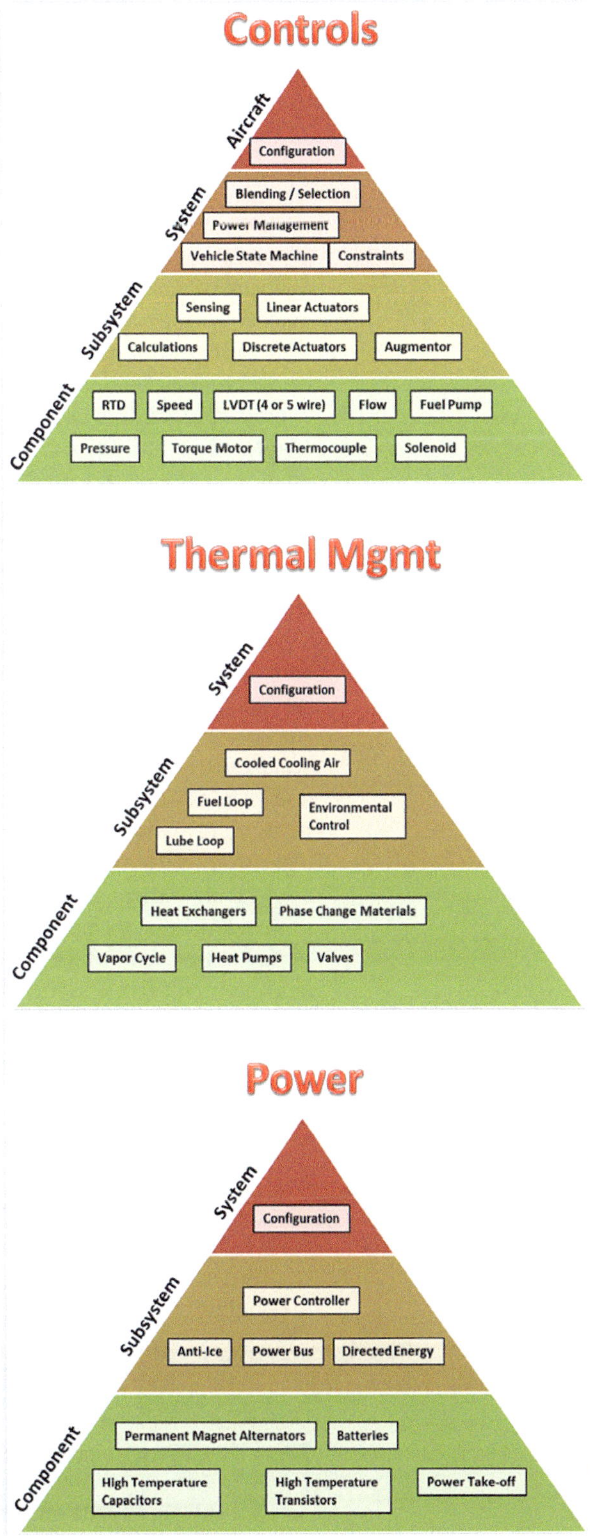

Hierarchical Adaptive Control

A hierarchical control architecture enables cooperative surveillance of the subsystems to cope with the nonlinear complex dynamical systems like an aircraft system. The hierarchical approach improves portability and the ability to distribute, as well as aligns software level with functional level and requirements. Integration, optimization, and management of dynamic couplings and nonlinear interactions of flight and propulsion systems as well as thermal and power management are too complex for real-time execution

FIGURE 7 Adaptive control is most versatile (From NASA 2004 [6]).

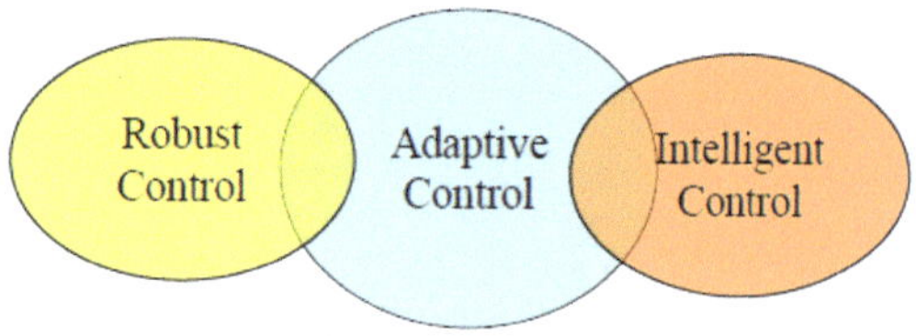

in onboard computational platforms that exist today. Hierarchical control architecture which combines continuously varying and discrete event architecture is the preferred approach for the development of future generation integrated control systems that will take the advantage of these interactions for mission enhancement. A hierarchical approach is to group and organize the distributed controls and processes into a logical hierarchy of multiple levels, so as to coordinate the local computation/control activities at each node to improve the overall system performance. Although the computing power of modern engine controllers continues to increase, the sophistication of the software being implemented on them continues to outpace the computational throughput capability. Integration and optimization further expands the computational need. In the future, algorithm growth will be driven by intelligent adaptive engine control modes. These modes will require additional sensors, actuation, and high-frequency local control. Multiple processor or coprocessing distributed architectures can be added to the network to offer opportunities to significantly increase throughput using currently certified electronics for use on gas turbine engine applications. With these alternative architectures come new challenges in ensuring that the resulting control implementations meet necessary functional and stability robustness requirements. The distributed networked controls is closely related to distributed computing which makes this approach very desirable for future aircraft expansion capability.

The hierarchical decomposition given in the software architecture connects each subcomponent by data buses and is naturally represented as a tree with internal smart nodes representing coordination components and external nodes representing termination components. The hierarchical architecture continues to show that it is flexible, scalable, maps easily to requirements, offers efficient parallel design by multiple designers, and provides ease of module reuse. Developing hierarchical control systems involves integrating many different types of components and subcomponents, including software algorithms, device drivers for hardware components, sensor data processing code, engine controls algorithms, and mathematical flight dynamics models written in more than one programming language. NASA has advanced the state of the art in adaptive controls as a design option that will provide enhanced stability and maneuverability margins for safe landing in adverse conditions [6]. A control system that continually changes itself to conform to changes in the "plant" (controlled system) and environment to maintain safe and satisfactory performance is labeled adaptive. Figure 7 illustrates how adaptive control systems incorporate elements from robust control theory and techniques as well as intelligent control concepts.

The hierarchical adaptive control scheme proposed in this paper improves flight safety [7, 8] (Figure 8). The FADEC computer takes commands from the pilot and flight control system to provide actuation commands to enable stable flight. It provides a safe/reliable mechanism for the control of the aircraft, including handling qualities, optimal engine performance, and logistics needs. Actuation systems play an important role for the engine/ aircraft/pilot/passenger, many are part of safety-critical systems.

Integrated Design of Reconfigurable Fault-Tolerant Control Systems

Active control of turbo-machinery requires closed-loop control of hydraulic, electromagnetic, or smart material actuation. Historically, hydraulic actuators are used in

FIGURE 8 Growth of adaptive control technologies.

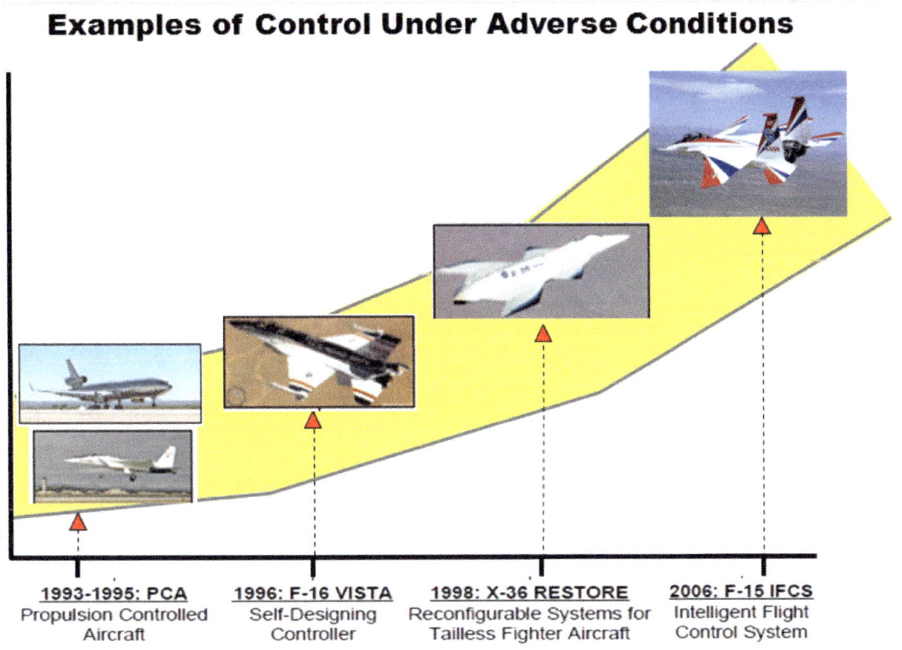

propulsion, but the trend is toward using more electric-motor-driven designs. New control technologies are currently being developed to provide alternatives to hydraulic actuation systems for various components of variable-cycle engines in addition to other needs for aircraft and engine controls. The controls for these actuation systems will require robust operation with the status available at all times to the FADEC and flight control to avert possible failures. A smart actuator will have integrated monitoring, control and health status software and electronics, eliminating hydraulic lines, brackets, and associated weight from the aircraft. A smart EMA is required to operate under vibration and high temperatures and need to respond very quickly. It should be fail-safe, closed or open, or locked into position on power or system failure. The actuator mechanism may be adapted particularly for position control and actuation of valve mechanisms, including those having a fail-safe feature. Self-contained smart EMAs could eliminate the hydraulic runs and reduce interface cabling. Failure detection onboard the actuator itself, which would greatly improve diagnostic and repair intervals at all maintenance levels, can increase their reliability. With proper use of a fiber optic data-bus interface, it can provide enhanced immunity to electromagnetic radiation including electromagnetic weapons.

Distributed Decentralized Control Architectures

The objective of distributed controls is to exploit the anticipated performance advantage of the centralized control while maintaining the scalability, local flexibility and customization, ease of implementation, and robustness of decentralized control [9]. The network that will be used to enable sensors, actuators, and controllers to communicate is considered as a data bus, and the transforming centers, equipped with supervisory elements, as data-bus terminals. The communication channel is characterized by means of well-known communication theory tools, appropriate signal structures, and the reception subsystem. Decentralization would have several advantages including improved survivability, code redundancy, and more efficient usage of computing resources.

This architecture facilitates faster and more accurate response by providing decentralized robust control to several selected nodes in each subsystem. At the local level, these decentralized robust controllers can be designed to quickly recover after large faults and disturbances in the system. The hierarchical configuration of distributed systems can achieve optimized system performance.

Zone-Based Autonomy and Authority

These concepts of decentralization and hierarchy in the control and management systems within the aircraft open up the opportunity to delegate authority to localized areas, or zones, throughout the aircraft and propulsion system. These zone-based controllers could be given autonomy, in some or all of the functional areas, to balance and regulate operating conditions, loads, data flows, and control commands. An architecture such as this would help to facilitate the differing demands and requirements at each level in the aircraft system hierarchy. For example, data refresh rates, control system response times, and physical system response times can all be ensured and handled by controllers that are physically local to a particular system segment or subsystem. Figure 9 gives a graphical depiction of the highest level of granularity within this structure.

It is important to point out that it is envisioned there will be a linkage between the fully integrated aircraft management systems and some form of data storage and processing entity. As the figure indicates, this could take form as a ground station, satellite, or really any off-board computing resource that could be considered to provide cloud-like services. Many of the diagnostic and prognostic health management functions would be managed and carried out on this remote computing resource. Using aircraft and engine data for trending, data mining, and machine learning is not a new concept; however, these activities have traditionally relied upon very limited sets of data that are only recovered post-flight by the maintainers. Having a real-time data link offers many opportunities for finer monitoring and easier access to data that might otherwise be lost in the event of a mishap. Zone 1 is mostly comprised of central computers A and B, along with some of the avionic subsystems that reside near this area of the aircraft. In the case of a manned aircraft, this would also include electronics for cockpit displays and controls and the output of the environmental control subsystem. The highest level control functions would reside in the central computers. Zone 1 would also provide the aircraft side of the data link to Zone 0, so it would be necessary to include communication/transceiver electronics. Zones 2-6 mark the fuselage, wing, and stabilator areas, respectively. Each zone has at least one data concentrator or local controller. The purpose of this component is to collect all of the sensor and actuator signals within a localized region. In general, this component may include communication protocols, signal conditioning, computing resources, control laws, and fault tolerance logic. The local controller/data concentrator would interface with the main backbone throughout the aircraft—bringing all of the local data back to Zone 1 to be incorporated into the higher level control functions. In the following section, Zone 7, which encapsulates the propulsion system in this partitioning, will be discussed in some more detail.

In summary, a zonal control architecture offers interesting opportunities for advanced aircraft that will provide reduced SFC, maintenance cost, improved structural life, enhanced safety, and improved fault tolerance. Enabling integration of engine/flight control, thermal management, aircraft state, and flight modes is key for this control architecture to achieve the expected benefits. Higher levels of integration of engine/flight

FIGURE 9 Distributed decentralized control architecture based on zonal autonomy.

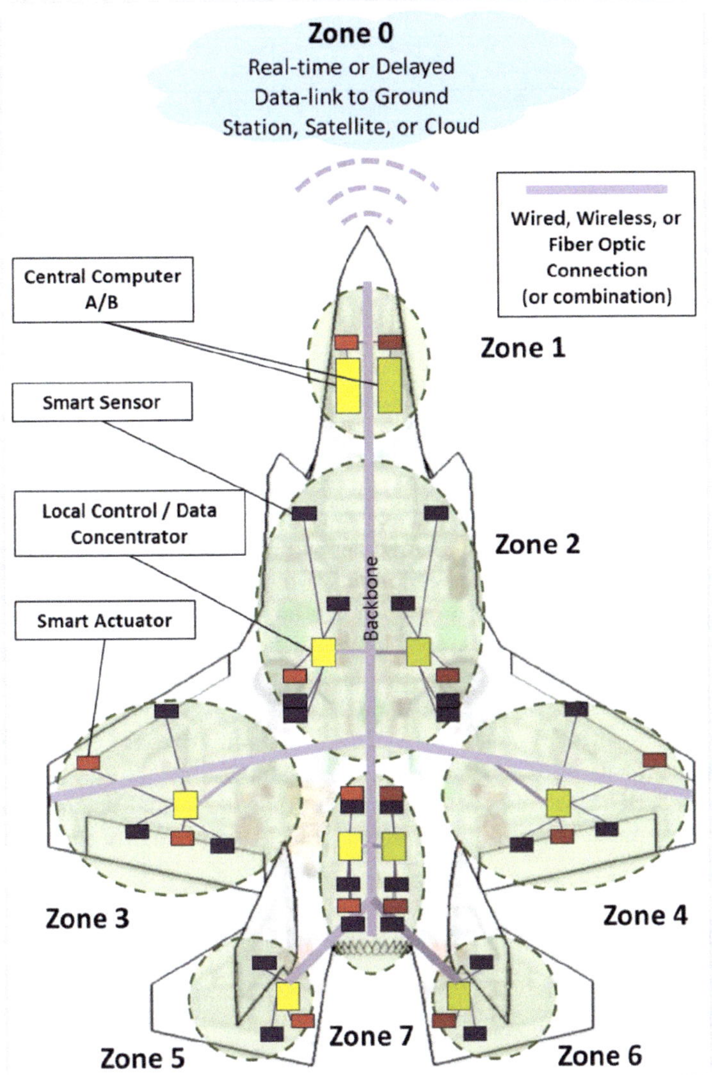

control can reduce fuel burn (SFC) through exploitation of advanced optimization, knowledge discovery, and PHM (fleet data) techniques. Enhanced safety is achieved in zonal control through its intrinsic graceful degradation properties of a distributed architecture which includes resilient communication, tailored response time, and controllability compared with a single point wired control architecture. Improvements in fault tolerance, damage tolerance, and maintainability can also be achieved with architectures based on localized autonomous functions. The ability to disable, reroute, and redirect control functions is significantly enhanced compared to current single point control systems.

Configurations for Distributed Decentralized Networks

Figure 9 shows a network configuration that can be thought of as a combination of star and bus topologies. The main backbones represent the bus portion of the network, while

TABLE 1 Summary of qualitative topology analysis for distributed engine controls [10]

	Ring	Star	Bus	Tree
Weight	(green)	(yellow)	(green)	(yellow)
Reliability	(red)	(yellow)	(red)	(red)
Maintainability	(red)	(yellow)	(yellow)	(red)
Msg TX Length	(red)	(green)	(green)	(green)

* Green markers indicate potential benefits, red markers indicate potential challenges, and yellow markers are neutral.

the data concentrators represent the center, or hub, of a star network. As drawn, this configuration may not necessarily be the best design, or what will eventually be implemented. The design of control and electronics network topologies can have a significant impact on system weight, reliability and redundancy, and maintainability [10]. Thus, it is important to analyze different types of topologies with respect to each of these factors. In [10], the network topology for a distributed decentralized engine controller is discussed. Strictly speaking, the discussion is applied to engine control components such as valves, pumps, actuators, and sensors (which may or may not have expanded functionality in the form of analog-to-digital conversion or signal processing) [10]. However, the discussion is relevant, also, to the present discussion of distributed decentralized aircraft control, power, and TMS. Table 1 summarizes the qualitative analysis for the ring, star, bus, and tree topologies for the engine control case.

To get a better sense of how these different topologies might be implemented on engine, for example, Figure 10 is included from [10]. This figure shows an optimal

FIGURE 10 Distributed engine control configurations with five data concentrators (a-c) and a baseline configuration (d) [10].

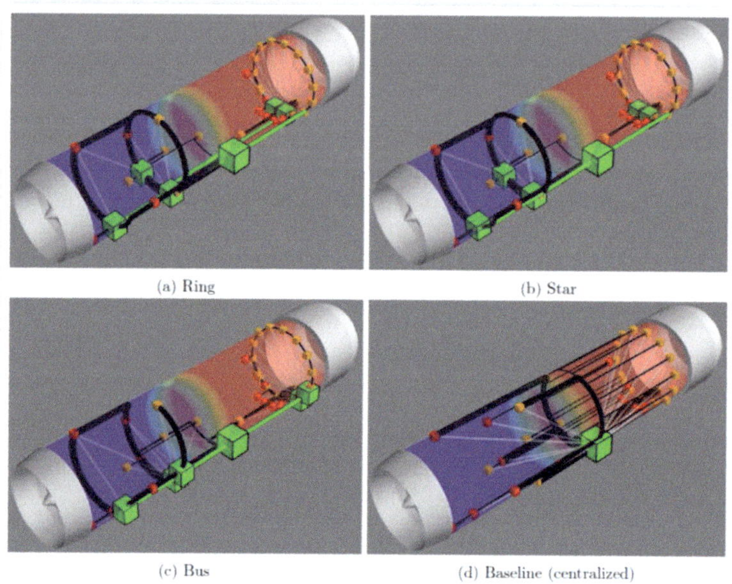

configuration of data concentrators in a ring, star, and bus physical topology alongside the baseline, centralized configuration (no data concentrators). The red, orange, and green blocks represent sensors, actuators, and data concentrators, respectively. Connecting each component is a black line whose thickness depends on the number of conductors in the cable that is represented. The main objective in the optimization of each configuration was to minimize total wiring length [10].

Based on the qualitative analysis Table 1 and the quantitative analysis (not reproduced here) based on Figure 10, it is unclear whether there is truly one basic network topology that is the best [10]. It is more likely that the design of a distributed decentralized aircraft management network will incorporate aspects of more than one topology in a sort of hybrid structure, as alluded to in Figure 9.

Practical Applications of Distributed Controls

Distributed controls enable local intelligence to be embedded or closely coupled with the subsystem or component hardware to monitor performance and communicate "condition" of the hardware. The practical applications of these control approaches support the implementation of condition-based system optimization. Using these approaches on future control systems will allow optimization of performance in addition to an increased understanding of the health of the equipment installed in the aircraft. The knowledge gained from this information allows the system to adapt to changes either within the hardware or by understanding the condition of other subsystems which influence the performance. The ability to distribute intelligence through the use of intelligent subsystems provides real-time feedback to the systems and allow reconfiguration to account for the degradation in a portion of the system.

As a simple example of this capability, consider a flight control system which is experiencing an electrical actuator failure. The actuator (intelligent) would communicate the condition of the degraded actuator (self-diagnosed). This would allow the flight actuation system to adjust other actuators to compensate for the failed or degraded hardware. This fault-tolerant reconfiguration of the system can be done in real time without operator or ground intervention. Distributed intelligence combined with advanced EHM controls algorithms could be used to sense subtle changes in the performance of the hardware prior to the failure and provide a prognostic which would allow maintenance prior to the actual impact on vehicle performance.

In the future, the ability to detect component degradation (intelligent subsystems) combined with an understanding of the actual environment (usage, temperature, vibration, etc.) the hardware has been exposed to can be used to understand the root cause of the degraded performance and to develop more reliable hardware solutions. Subsystems suppliers can use this application knowledge to provide improved reliability and better understand environmental issues which impact performance and lifing.

The proposed approach is the new generation of total aircraft system integration for the optimized controls, thermal, and power in collaboration with OEMs, airframers, and supporting industries. It is designed to maximize plant efficiency and reliability through distributed control, integration, and optimization of the entire plant. The new approach offers a simple, scalable, seamless, and secure comprehensive solution. The significant opportunities created by the proposed distributed control architectures will require changes in the aerospace industry to overcome the technical and business challenges that result as propulsion, thermal, and electrical solutions become highly integrated. The successful AFRL INVENT program focused on platform system engineering and model-based design to overcome the technical challenges by working the platform solutions with an integrated IPT which included propulsion, electric, and thermal disciplines. These teams, led by the airframers, used a strong reliance on the engine manufacturers to bring integrated power solutions (variable-cycle propulsion, power and thermal systems) forward to the architecture studies.

Future programs will benefit from the modeling and simulation standards like the AFRL-developed Modeling Requirement and Implementation Plan (MRIP) which allowed both mission and segment (high fidelity) simulations to be performed at the platform level. Standardized controls communications protocol and messages will support and accelerate the development and deployment of these future systems.

Certification/Qualification

A significant but not insurmountable challenge exists with these new architectures and concepts in the demonstration of flight safety and reliability. As the level of integration of multiple subsystems and components to accomplish this optimized system is currently unprecedented, new approaches and new processes will be required to demonstrate to certification authorities the safe and reliable operation of the air vehicle. Key to this will be extensive fault hazard assessments and innovative fault detection and accommodation strategies to adapt and accommodate to equipment health/status in real time. Current certification approaches stress individual, nonintegrated system reliability, and fault accommodation. These new distributed and integrated architectures will require new approaches and certification processes to ensure safe and robust integrated vehicle operation.

Systematic Propulsion System Optimization

This paper proposes an integrated control, energy, and thermal management strategy for an optimized propulsion system. Clearly, integrated design by optimization, optimization process, and optimization criteria have to be extensively defined first. The proposed approach emphasizes the prime importance of mission profile issues in zone-based distributed system and, more generally, of environmental variables for aircraft-engine level and subsystem. This supervisory hierarchical distributed control strategy is developed based on the equivalent consumption minimization strategy to optimize turbine engine, TMS, power, and weapon system and for the entire system. The communication data bus, distributed control, smart nodes, zone-based autonomy, and authority along with hierarchical controls allow engine operating schedules to be adjusted in flight on the basis of adaptive optimization algorithms which identify engine component performance variations due to manufacturing, wear, and damage.

A quantification for optimization has to be made for performance benefits accruing for in-flight optimization. The performance benefits of adaptive in-flight optimization can be achieved via modifying control gains and model parameters based on high fidelity estimates of high-level performance parameters. For both real-time and offline optimization strategies, several analytical optimization methods such as Pontryagin's minimum principle [11], static optimization, dynamic programming (DP), quadratic programming (QP), model predictive control (MPC), linear programming, and Hamilton-Jacobi-Bellman equation are available for control optimization. Reference [12] classifies control strategies as either optimal or suboptimal. Global optimization methods, such as DP, can be combined with optimizations performed at the local level for reliable solutions. The other categories include heuristic control, rule-based power management control, and equivalent consumption minimization strategy [13, 14].

Summary/Conclusions

In the past, a lack of close integration within and between various aircraft systems and subsystems lead to worse case conditions, when combined, resulted in large operation margins for legacy aircrafts. Propulsion played a role only in generating thrust, power systems were mostly for auxiliary purposes, and thermal management was only necessary under extreme areas of the operational envelope. Reliability and robustness with respect to accommodating faults and equipment health status of the integrated system, real time, must not be overlooked in order to ensure safe operation.

Going forward, amidst the quest for higher performance, greater control, longer life, lower cost, and unrestricted operation, close integration between adaptive propulsion, controls, power, thermal management, diagnostic, and prognostic systems becomes paramount. Each of these areas will continue to become more and more important to the operation of an aircraft across all areas of the flight envelope. It is important to note that, historically, many propulsion systems, OEMs, or system integrators who often have little or no interest in the energy use of their equipment addressed the optimization. This is an area where legislation, in the form of a requirement for an energy declaration on machinery and systems, could encourage greater consideration to be given to LCCs. It is envisioned that, from a system architecture standpoint, a hierarchical structure is necessary to facilitate control and communication both vertically, between high-level control functions and lower level components, and horizontally between different functional areas.

This paper proposes a decentralized distributed network structure for enabling these types of interconnections. Autonomy in balancing and regulating operating conditions, loads, data flows, and control commands is partitioned and delegated to local zones in this scheme. The movement toward a truly integrated aircraft/vehicle management and control system points to a need for improved software and algorithms, new controls hardware (sensors, actuators, communication network), and improved power and thermal management elements. System integration may allow these margins to be reduced, resulting in higher thrust, better safety fuel efficiency, and reliability.

Acknowledgments

The authors would like to acknowledge the contributions of those who made this paper and its contents possible: John Nairus, Scott Burns, Ken Semega, and Anhtuan Ngo from Air Force Research Laboratory. Any opinions, findings, and conclusions or recommendations expressed in this material are those of the author(s) and do not necessarily reflect the views of the USAF or Air Force Research Laboratory.

References

1. Stewart, J., Burcham, F. Jr., and Gatlin, D., "Flight-Determined Benefits of Integrated Flight-Propulsion Control Systems," NASA Technical Memorandum 4393, 1992.
2. Jacobs, A., "Reconfigurable Fault Tolerance for Space Systems," Dissertation, University of Florida, 2013, http://ufdcimages.uflib.ufl.edu/UF/E0/04/53/37/00001/JACOBS_A.pdf.
3. Integrated Propulsion, Power and Thermal (INPPAT) for Next Generation Capability, August 2013, John Nairus, Chief Engineer, Power & Control Division, Aerospace Systems Directorate.
4. NASA Presentation Contributed from DECWG, "Distributed Engine Control, NASA Fundamental Aeronautics Program Subsonic Fixed Wing Project," *2009 Propulsion Controls and Diagnostics Workshop*, December 8-10, 2009.
5. Dalton, J., Behbahani, A.R. et al., "Vision of an Integrated Modeling, Simulation, and Analysis Framework and Hardware-Test Environment for Propulsion, Thermal Management and Power," *43rd AIAA/ASME/SAE/ASEE Joint Propulsion Conference & Exhibit*, AIAA 2007-5711, 2007.
6. Aviation Safety Program Integrated Resilient Aircraft Control Project, Project Overview and Status, Totah Joseph, Principal Investigator, Krishnakumar Kalmanje Dr., Project Scientist, Viken Sally, Project Manager; http://www.eng.morgan.edu/~cibac/events/Day1/9-IRAC%20Overview%20(Totah)pdf, 2007.
7. Idan, M., Johnson, M.D., and Calise, A.J., "A Hierarchical Approach to Adaptive Control for Improved Flight Safety," *AIAA Journal of Guidance, Control and Dynamics* 25, no. 6 (2002): 1012-1020.
8. Nguyen, N. and Jacklin, S., "Neural Net Adaptive Flight Control Stability, Verification and Validation Challenges, and Future Research," *Workshop on "Applications of Neural Networks in High Assurance Systems," International Joint Conference on Neural Networks*, 2007.
9. Behbahani, A., Culley, D., Carpenter, S., Mailander, B. et al., "Status, Vision, and Challenges of an Intelligent Distributed Engine Control Architecture," SAE Technical Paper 2007-01-3859, 2007, doi:10.4271/2007-01-3859.
10. Von Moll, A. and Behbahani, A.R., "Comparison of Communication Architectures and Network Topologies for Distributed Propulsion Controls," *59th International Instrumentation Symposium*, Cleveland, 2013.
11. Yuan, Z., Teng, L., Fengchun, S., and Peng, H., "Comparative Study of Dynamic Programming and Pontryagin's Minimum Principle on Energy Management for a Parallel Hybrid Electric Vehicle," *Energies* 6 (2013): 2305-2318.
12. Hofman, T., Steinbuch, M., van Druten, R.M., and Serrarens, A.F.A., "Rule-Based Energy Management Strategies for Hybrid Vehicles," *Int. J. Electr. Hybrid Veh.* 1 (2007): 71-79.
13. Wang, X., He, H., Sun, F., Sun, X. et al., "Comparative Study on Different Energy Management Strategies for Plug-In Hybrid Electric Vehicles," *Energies* 6 (2013): 5656-5675.
14. Xu, Q., Cui, S., Song, L., and Zhang, Q., "Research on the Power Management Strategy of Hybrid Electric Vehicles Based on Electric Variable Transmissions," *Energies* 7 (2014): 934-960.

Improved Engine Health Monitoring Using Full Flight Data and Companion Engine Information

Allan J. Volponi and Liang Tang
Pratt & Whitney

Engine module performance trending and engine system anomaly detection and identification are core capabilities for any engine Condition-Based Maintenance (CBM) system. The genesis of on-condition monitoring can be traced back nearly four decades, and a methodology known as Gas Path Analysis (GPA) has played a pivotal role in its evolution. GPA is a general method that assesses and quantifies changes in the underlying performance of the major modules of the engine (compressors and turbines) which directly affect performance changes of interest such as fuel consumption, power availability, compressor surge margins, and the like. This approach has the added benefit in that it enables anomaly detection and identification of many engine system accessory faults (e.g., variable stator vanes, handling and customer bleeds, sensor biases and drift). Legacy GPA has been confined to off-board analysis of *snapshot* data averaged over a stable flight conditions when the engine is in steady-state operation. This discrete data point approach, while fairly accurate and repeatable, comes with a price tag which is a time latency to detect (and subsequently isolate) a faulty condition. In this paper, we explore the use of streaming full flight data which includes both transient and steady-state operation. This type of data stream, when properly processed, allows faster anomaly detection, credible fault persistency checks, and timely fault identification. This paper outlines the use of companion engine data to achieve a cross-wing comparative diagnostic method.

Introduction

For nearly four decades, GPA has provided a valuable tool for tracking long-term performance deterioration of the major modules of a turbofan gas turbine engine in support of the on-condition maintenance philosophy that has evolved during that period. As such, it has been useful in assessing general engine health, in support of fleet-wide engine trending and depot maintenance logistics, as airlines transitioned from *scheduled* maintenance based on hours flown to a more Condition-Based Maintenance (CBM) operation based on individual engine health. The original formulations have evolved and expanded to include a diagnostic capability for the Fault Detection and Isolation (FDI) of engine subsystem components whose deterioration or failure result in observable performance changes in the engine's gas path and as such has had an impact on line maintenance practices as well. These subsystems include engine bleed systems, actuated variable geometry, turbine clearance control, instrumentation, and the like. The literature abounds with descriptions of engine monitoring data capture systems, applicable algorithms for the analysis of the gathered data, and the perceived cost benefit of these practices. The interested reader can find a summary discourse and extensive bibliography on the subject in references [1, 2, 3], and as such, we will not repeat a detailed description of legacy methods in this paper. However, to familiarize the reader, who may be unacquainted with these methods, we will provide a short summary of the salient concepts in order to motivate the central theme of this paper, which is the extension of these methods to leverage full flight data in the emerging world of *big data*.

The format of the paper begins with a summary section on historical perspective, the benefits derived from considering full flight data, the technical challenges imposed, the motivation for utilizing companion engine data, and finally a particular approach that utilizes this *cross-wing* strategy to mitigate these challenges. Simulation studies will be discussed in detail along with pertinent metrics to measure the efficacy of this approach.

Historical Perspective

Engine diagnostic practices are almost as old as the gas turbine itself. Monitoring and analysis methods have progressed in sophistication over the past decades as the gas turbine evolved in form and complexity. Beginning with mainly empirical methods centered on monitoring the mechanical integrity of the machine, the evolution of engine diagnostics has benefited from advances in sensing, electronic monitoring devices, increased fidelity in engine modeling, and analytical methods. The principal motivation in this development is, not surprisingly, cost. The ever-increasing cost of fuel, engine prices, spare parts, maintenance, and overhaul all contribute to the cost of an engine over its entire life cycle. Prognostics and diagnostics can be viewed as a means to mitigate risk in decisions that impact operational integrity. This can have a profound impact on safety, such as In-Flight Shutdowns (IFSD) as well as the economic impact caused by Unscheduled Engine Removals (UERs), part life, maintenance and overhaul, and the overall logistics of maintaining an aircraft fleet.

Engine Health Monitoring began in earnest with the advent of commercial high bypass turbofan engines introduced in the 1970s. On-board data collection systems initially focused on collecting data from the existing instrumentation available at the time which was primarily gas path sensors (speeds, temperatures, and pressures) that were resident on the machine for engine control purposes. The development of GPA began in earnest as commercial airlines and engine manufacturers used this data to trend engine performance. The data used for this purpose was stable aircraft cruise data that would be averaged (over a short time window) to produce a *snapshot* of various engine and flight parameters for subsequent analysis and trending off board.

Classical GPA was targeted to track long-term engine module performance degradation as measured by thermodynamic changes in efficiency and flow parameter for each major module (compressors and turbines). It was later modified to accommodate additional functionality such as fault event detection and isolation to address abrupt events and failures such as Foreign Object Damage (FOD), Domestic Object Damage (DOD), bleed leaks and failures, variable geometry anomalies, actuator and instrumentation faults, and the like. This legacy gas path FDI relied on the ability to recognize abrupt shifts in steady-state snapshot cruise data. Parameter Δ*s* from a (fleet) normalized reference model provided the raw ingredients to enable both the detection and identification process. Long- and short-term averages of the parameter Δ*s* were compared to determine if a divergence was present (indicating that an abrupt shift in the parameter had occurred) and a *persistency* check was made to determine that the shift was real and not due to an operational excursion. If a persistent shift was observed, parameter ΔΔ*s* would be calculated between the initial (pre-shift) Δ value and its final (shifted) Δ value to form a *signature* across all dependent measured gas path parameters which in turn could be used to isolate the underlying fault through a pattern matching strategy under the hypothesis of a *single fault* scenario. The details of this process can be found in [4].

Over the subsequent four decades since its inception, GPA has been the topic of much research which has extended its capabilities from its early formulation. Its success has contributed to its longevity, and it remains as one of the central techniques supporting health management to this day.

Full Flight Data

Since the inception of engine condition monitoring four decades ago, the on-board monitoring hardware has dramatically improved in both capacity and capability. With enhanced multiprocessor CPU throughput and massive storage capacity at hand, *both* on-board analyses of engine data is within reach as well as the ability to provide streaming data for off-board processing to ultimately replace the legacy snapshot analyses in current practice. The latter path is part of the *big data* phenomena we hear so much about. The choice of deploying on board or supporting the *big data* approach has many considerations, for example, implementation complexities, certification issues, data management issues, analysis maturation, and life-cycle cost, to name a few. What is important, from the authors' perspective, is that the analyses methodologies created should be configured to be as generic as possible to allow either on-board or off-board implementation with minimal change. If we restrict ourselves, for the purpose of this paper, to gas path data, what does a full flight perspective offer us beyond what the legacy snapshot approach already provides, given its stated success and longevity? The simple answer is, *time*, that is, a *reduction in the time latency* to provide credible FDI with minimal false alarms and missed detections. Figure 1 illustrates the problem faced with *snapshot* data. The area in the plot that is shaded in pink indicates the observational emergence of an event, that is, rapid deterioration of a module or failure (full or partial) of an engine subsystem (such as a bleed) or instrumentation fault. The portions of the plot shaded in yellow signify the effect on the measurement delta due to long-term (gradual) component degradation. Recall that each point in this plot would signify a single (averaged) snapshot data point taken from steady-state cruise flight and the time axis is measured in flights (i.e., days rather than seconds). Thus, such a plot would gradually emerge with each subsequent

FIGURE 1 Snapshot parameter Δ trending and event detection.

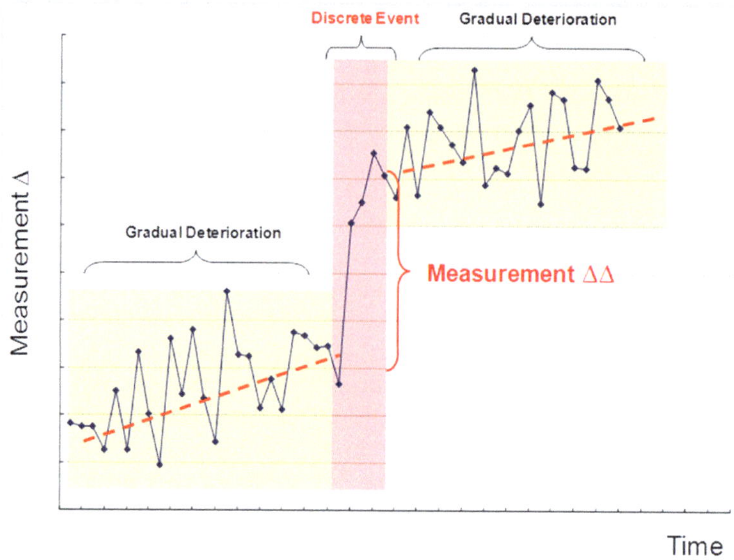

flight and that a decision would have to be made as each data point was processed. In this hypothetical plot, there are four data points depicted in the pink area. The event detection problem must weigh the consequences of a false alarm versus the consequences of waiting too long to be assured that an event has actually occurred versus a set of rouge points (outliers). This is the general problem of determining *persistency*. The detection methodology employed must minimize false alarms due to outliers while reducing the risk of catastrophic failure which might occur on subsequent flights if the underlying fault was critical and persistency was not established fast enough to allow its determination.

The use of full flight (streaming) data offers the potential for many more data points (within the current flight) to establish a *persistent* shift and launch a single fault isolation procedure immediately. The plurality and fusion of such analyses will mitigate the risk of false alarms and missed detections as well as reducing the potential for collateral (i.e., secondary) damage that might occur if the FDI process were allowed to span several flights, as is the case with *snapshot* data usage.

Technical Challenges

The use of streaming data offers a number of technical challenges to afford the same FDI functionality provided with legacy snapshot approach. The ability to perform this same functionality on board with streaming data, both steady state and transient, requires a number of changes in basic strategy. The first centers on the *reference* model (used to generate the measurement parameter Δ*s*) and the long-term performance tracking. The traditional discrete Kalman Filter (KF) typically used in the snapshot analysis must be recast as a dynamic tracking filter in closed-loop operation with a transient (physics) model of the engine. The latter can be a piecewise linear State Variable Model (SVM) or a more sophisticated nonlinear component level model and the KF can be derived first from an SVM and later applied to a nonlinear model if required. The KF states are extended to contain health parameters (efficiencies and flow capacities) that act as *tuners* to track health deterioration of the engine over the long term. Unfortunately, these tuners can be corrupted if there is any modeling error, instrumentation bias, and engine-to-engine variation. This mismatch between the specific monitored engine and physics model is apparent in the data and is most severe during transient operation. This necessitates that a normalization of the model be completed to remove these inconsistencies and allow the KF to track health deterioration in an uncorrupted manner. As we will see in a moment, this is also a necessary (but not sufficient) condition to enable proper anomaly detection and isolation of abrupt faults. One strategy for accomplishing this is to construct an empirically derived difference model between the monitored engine and the model and then use this to initialize the physics model [5, 6]. This will remove the steady-state and transient errors and allow the tuners to track health over the life cycle of the engine. This stabilization between the engine observations and the model reference afforded by the normalization permits more accurate anomaly detection and reduces false alarms. Other methodologies can be found in [7, 8].

Event (or anomaly) detection and subsequent fault identification is another capability to address in the streaming data environment. This issue is just as insidious as the model normalization task in the technical difficulties that it presents. Fault events (of the same type that are addressed in the snapshot analysis world) must be detected and isolated during transient conditions as well as steady-state operation. Extensions of the methods typically applied in steady state, seem to perform adequately enough for near-steady-state conditions when streaming data are applied. Transients, however, do not fare as well. The legacy detection methods do not work well and identification methods miss the target more often than not. This paper describes an approach that addresses these issues and will allow extensions of these same methods to be utilized in situations where there are two (or more) engines present on the airframe.

Companion Engine Data

The basic idea is to use information available from the companion engine(s) through a cross-wing comparison. Since companion engines may have different performance levels (FADEC plug class), may operate at slightly different power levels (due to performance or other operational considerations, e.g., airframe control trim, etc.), and may have different instrumentation errors or bias anomalies, a direct comparison of engine parameter values is not possible (or desired). The inclusion of an engine model for each engine, properly tuned or initialized to represent the individual engines, would provide a means to mitigate these concerns. Residual differences (Δs) for each gas path parameter relative to its individual normalized model remove the performance differences, instrumentation error differences, plug class differences, power level differences, etc., and permit a cross-wing comparison of the individual parameter Δs to be made for anomaly detection purposes. In essence, the companion engine with its attendant normalized model becomes the *reference* for the monitored engine under analysis. Deviations due to environmental changes, flight maneuvers, and long-term deterioration will effectively cancel out in the cross-wing comparison. The latter item assumes that after initial calibration and model normalizations, both engines will essentially deteriorate at *approximately* the same rate.

This assumption is made on the basis that both engines are exposed to the same environment, flight maneuvers, and power excursions. In this way deviations not common to both engines will become visible in the cross-wing comparison. In the next section, we explore the cross-wing analysis architecture and describe its attendant algorithms.

Cross-Wing Analysis Architecture

Before introducing the proposed methodology, it is important to understand the scope under which it is to be employed. The goal is to support engine health management decisions and enable faster and more accurate FDI to support maintenance activities. Although the process and algorithms to be discussed can be executed on board, it is not intended, at least at this stage of initial development, to offer real-time cockpit annunciation or

FIGURE 2 Cross-wing analysis architecture.

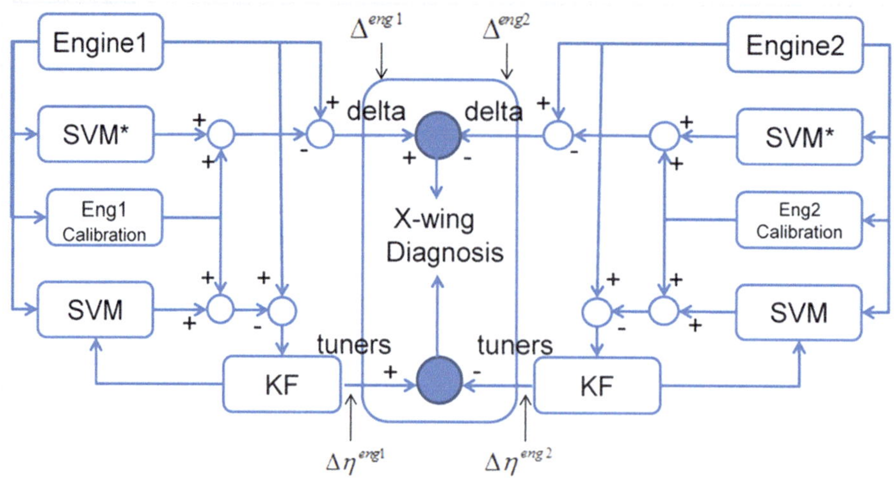

accommodation. As such, the final analysis alerting will occur at the *end of flight* to take full advantage of the information fusion afforded by considering full flight data.

The proposed architecture for this cross-wing comparative FDI is depicted in Figure 2. For simplicity, we have depicted a two-engine aircraft configuration. It should be noted, however, that this can be generalized to multi-engine (>2) aircraft configurations by applying this framework pairwise across remaining engines.

The architecture is clearly symmetric. The entity labeled SVM is State Variable Model, and there are four *identical* instances of this model depicted in the figure. The asterisk (*) that appears on two of them is only intended to distinguish it from its neighbor (for a given engine, say, Eng1) in that while identical it may be operating in a different state since it does *not* have a feedback from the attendant KF estimator. The blocks labeled *Calibration* provide a normalization that is *unique* to each engine in order to align the model more closely to the engine at installation. It is depicted as a bias adder in the figure as a generality to allow any form of model (linear or nonlinear) to be substituted for the SVM without need for changing the architecture or the FDI process to be described herein.

The FDI process, in overview, can be explained as follows. The cross-wing comparison $\Delta\Delta s$ (denoted $\Delta\Delta^{\times} = \Delta^{eng1} - \Delta^{eng2}$) for two *calibrated* engines would be (on the average) zero during any operation. If one of the engines suffers a rapid deterioration due to FOD or DOD or is the victim of an abrupt engine subsystem or instrumentation failure (full or partial), the cross-wing $\Delta\Delta^{\times}$ will diverge in consequence. Comparing the $\Delta\Delta^{\times}$ to an appropriately chosen threshold and applying a persistency check criteria will flag the anomalous behavior. This provides a detection method across the full flight regime. It does not, however, determine which of the two engines is in fault. To determine this, a second identical normalized model (per engine), fitted with a KF observer (used for long-term performance deterioration tracking), is required. The faulted engine's KF *tuners* will react (i.e., diverge) as a result of the fault event, whereas the non-faulted engine's KF tuners will not. This allows the identification of which engine has sustained the faulty condition. Furthermore, the cross-wing $\Delta\Delta^{\times}$ across all measured engine parameters provide a *signature* for subsequent fault isolation. In the next sections, we will explore some of the details inherent in this architecture.

Engine Model Normalization

It should be no surprise that the engine model plays a pivotal role in any model-based FDI method. It forms the fundamental reference framework representing the nominal state from which (observed) parameter deviations can be quantified as the engine degrades and/or faults occur. Variation in the form of production/overhaul engine-to-engine differences, modeling infidelities/simplifications, instrumentation biases, etc., is ever present and must be accounted for in any model-based diagnostic/prognostic approach. Minimizing this variation is critical to accurate engine health monitoring analyses. A fundamental practice to mitigate this variation is engine model *calibration* (or *normalization*). This is the process of aligning the model to the actual engine under observation for diagnostic systems that perform, at least some of its functions, on board in a quasi-real-time environment. This alignment can be achieved in a variety of ways through adjustments to the model in the form of corrections or normalization factors whose net effect is to reduce the differences between model output and actual (sensed) observations. In any event, the efficacy of the calibration method and process will depend on the amount (and type) of data available, time allotted to accomplish the calibration, robustness of the process (i.e., can it be applied generically to any model formulation), and, of course, the level of error that can be tolerated in the calibrated model for the specific application at hand. It is outside the scope of this paper to describe, in detail, a specific calibration procedure; rather, we will explore some of the high level properties and consequences of a calibration within the diagnostic process.

QUASI-STEADY-STATE

Aero engines exhibit a wide-ranging level of dynamic behavior. It is fair to say, from a model error perspective, that steady-state operation is undoubtedly an easier mode of operation to model than the almost infinite transient maneuvers that an engine might experience during flight. As a consequence, model calibration for *steady-state* operation is a far easier task than a model calibration that encompasses all possible engine transients. As a matter of practicality, a *reasonable* model calibration would capture something in between these two extremes. We will use the term *Quasi-Steady-State* (*QSS*) to designate the *collection of conditions for which an acceptable alignment* (*i.e., calibration*) *can be made*. The term acceptable, will depend on many factors, including model type, maximum tolerable error, time to calibrate, and other elements germane to our particular application. We will assume, without loss of generality, that *QSS* consists of a collection of engine operating and flight regimes that contains steady-state operation and some subset of transient behavior. Whatever the form the calibration process takes, we will assume that the following two (fairly obvious) properties apply: (a) If A is an operating regime that can be calibrated, then any subset of A can be calibrated, and (b) the union of calibrated regimes can also be calibrated. With those assumptions, it is clear that *QSS* will consist of a union of calibrated operating regimes (A_k).

The form of A_k will vary according to the specifics of the calibration process (and model configuration) in use, but in general will be a function of ambient conditions, flight conditions, and engine operating parameters (e.g., power level, variable geometry, internal bleed status, etc.). It is apparent from these two properties that calibration can be a sequential process which can occur incrementally over a number of flights.

What is important, for FDI, is the ability to recognize whether data being processed (or collected) is within a particular regime that is contained in the *QSS* collection. This question can only be answered if we know the particular calibration process being applied. We will defer this question for discussion in the Simulation Verification section later in this paper and first explore some consequences of having an already defined *QSS* calibration regime and how it is exploited in performing an engine FDI.

FDI Process

Before we can identify a fault, we must know that one has actually occurred, that is, we must first *detect* the presence of, or consequence of, a fault. In this context, we assume a fault to be some part of the engine system or one of its subsystems which have ceased to function as intended. This may be a total failure or a partial failure of the system or subsystem element which (typically) occurs over a short period of time, thus distinguishing itself from normally expected long-term degradation or deterioration. To be detectable, it must also affect some set of parameters that we have available to us to observe, that is, sense. Since failed elements do not repair themselves, their effect on what we can observe must also be *persistent*. This latter requirement lessens the likelihood of a misdetection due to spurious operational events and noise. Lack of model calibration can also lead to misdetection if the model infidelity during certain operational maneuvers provides sufficient departure of model response to actual observation over a finite period of time which can confuse the situation with a false sense of persistency. For these reasons, *detection should only be performed when operating within QSS*.

When the model is operating within the *QSS* regime, the tuners, $\Delta\eta$, provide adjustments to the model to reflect the natural deterioration of the engine over a long period of time. However, when operating outside the QSS regime, the tuners become a mathematical artifact to force the parameter residuals to zero. Similarly, when a fault occurs, the tuners will likewise respond in order to force the residuals back to zero. Thus if we are operating within *QSS*, the tuner values can be used to test for the presence of an unknown disturbance, which if persistent, indicates the presence of a (yet to be identified) fault. This is accomplished by determining if the *cross-wing tuner* comparison $\Delta\Delta\eta^{\times} = \Delta\eta^{eng1} - \Delta\eta^{eng2}$ (see Figure 2) has shifted persistently (in *QSS*). If it has, then either Engine 1 or Engine 2 has sustained a fault. To determine which engine, we need only test to see which of $\Delta\eta^{eng1}$ or $\Delta\eta^{eng2}$ has shifted. We set a fault flag *f* as follows:

$$f_i = \begin{cases} 1 & \Delta\eta^{engi} \text{ has shifted} \\ 0 & \text{otherwise} \end{cases} \tag{1}$$

We can then *adjust* the polarity of the cross-wing parameter signature ($\Delta\Delta^{\times}$) as follows:

$$\Delta\Delta^{\times} \leftarrow sign(f_1 - f_2)\Delta\Delta^{\times} \tag{2}$$

Equation (2) now provides the requisite signature (for the faulted engine) to be used in the isolation process. Following the procedure outlined in [4], a *single fault* hypothesis pattern match is performed against a set of predefined (single) faults, and the one admitting the least measurement error* is chosen. This isolation procedure is performed at every time step that the engine remains in *QSS* after fault detection is established. This will produce a multiplicity of individual isolation results, that is, one for each time step that the engine operates in *QSS* after detection. As one might suspect, it is likely that they will *not* all be the same diagnosis due to noise, calibration deficiencies, and the like. Since the analyses are intended for maintenance advisory, it is permissible to wait until the end of flight to make a final diagnostic determination. In this manner, some form of information fusion can be adopted to minimize risk of false assessment. The tactic used in this development was to employ a simple *majority vote selection*. As we will see in the next section, this simple fusion strategy proved reasonably successful.

One final comment on this process is in order. The process described in the preceding paragraphs does *not* assume that the actual fault *occurred* while the engine was in *QSS*. It may or may not be clear to the reader that the detection scheme, as described, can be reliably achieved if the fault occurred during an earlier *transient* maneuver of the flight. Moreover, it may not be clear that the $\Delta\Delta^{\times}$ calculated at a later time in *QSS*, indeed, provides a representative *signature* for the fault that occurred. We cannot offer a mathematical proof that it does; however, we will demonstrate its effectiveness through exhaustive simulation studies in the next section.

* This measurement error is the comparison between $\Delta\Delta^{\times}$ from Equation 2 and the *anticipated* $\Delta\Delta$ that would resul3t given the single fault being tested.

Simulation Verification

In this and subsequent sections of the paper, we offer an initial verification of the methodology described above. As with most Prognosis and Health Monitoring and FDI developments, this will be accomplished through simulations where we can control the insertion of a wide variety of engine system faults and measure the algorithm's effectiveness. Ultimate validation using actual engine data is a long process that occurs afterward during the method's maturation. While gas turbines are very complex machines, they are also very reliable and actual faulted engine data is opportunistic at best since they are also very expensive machines which generally prohibit seeded fault testing to be realized to any great degree. Therefore it is important for the verification process that the simulations be as exhaustive and realistic as possible, in order to flush out any inherent deficiencies in the methodology. To this end, the NASA C-MAPSS40K engine model was chosen for this simulation study. C-MAPSS40k is a Matlab/Simulink-based generic closed-loop, nonlinear, component level engine simulation modeled after a twin spool, high bypass turbofan engine in the 40,000 lb takeoff thrust [9, 10]. There are two primary reasons for selecting C-MAPSS40K. First, engine faults can be easily simulated. A wide range of fault modes including major module performance faults, subsystem (actuator) faults, and sensor instrumentation faults have been simulated in this study to achieve a comprehensive performance assessment without restricting to a narrow selection of faults types. This wouldn't be possible if using real engine data. The fault data generated by C-MAPSS40K contains realistic transient behaviors due to the reaction of the built-in nominal engine controller. Secondly, C-MAPSS40K does not contain proprietary information and has become a popular benchmarking platform in the Engine Health Monitoring (EHM) community in recent years [11]. In addition, the C-MAPSS40k software includes an atmospheric model that allows the engine to be operated at various altitudes, speeds, and ambient temperatures. This has allowed us to drive the simulations with actual Flight Operations Quality Assurance (FOQA) data to mimic realistic engine operations across flight regimes without restricting to a particular operating condition (such as sea-level static).

FOQA Data

FOQA, also known as flight data monitoring or flight data analysis, is a voluntary safety program in which commercial airlines share de-identified flight data with the FAA in an joint effort to find new ways to improve flight safety and increase overall operational efficiency [12]. The FOQA data used for this study contain 1 Hz sample data for altitude, Mach number, total air temperature, and power lever angles (PLAs) for both engines (on the same aircraft) from a total of 69 revenue service flights. A representative FOQA flight (#2 in the data set) is plotted in Figure 3. It should be noted that the PLAs for the two engines are slightly different and exhibit noticeable variations during climb and the two cruise flight segments at around 26,000 and 9900 ft. This is apparently a short haul flight. More FOQA flight profiles are shown in Figure A.1 in Appendix. After some necessary unit conversions, all the 69 FOQA flights were used to drive C-MAPSS40K simulations for this study.

Sensor Noise Simulation

Realistic sensor noise was added to the C-MAPSS40K simulations for all FOQA flights as shown in Table 1 below.

FIGURE 3 Representative FOQA flight data (Flight #2 in the data set).

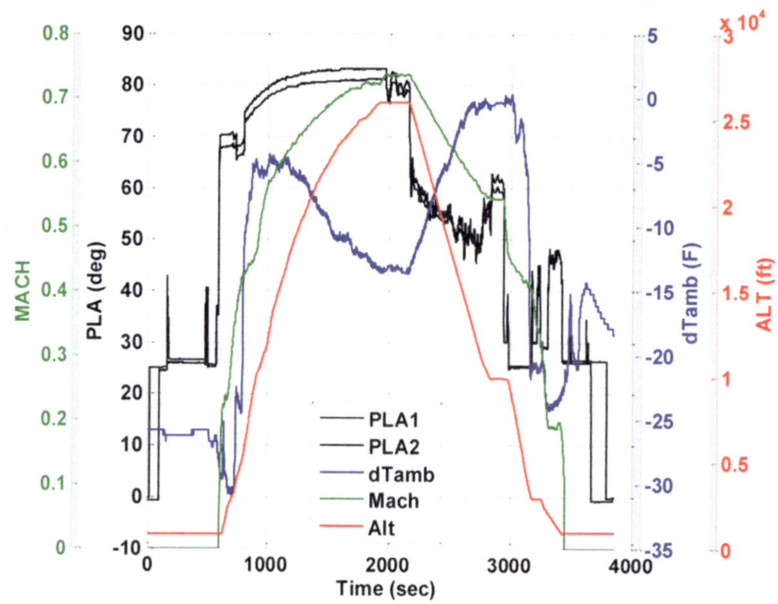

TABLE 1 C-MAPSS40K sensor descriptions and noise simulation

Sensor	Description	Noise std. deviation
N1	Fan speed	0.3 RPM
N2	Core speed	1.5 RPM
P2	Fan inlet pressure	0.2 psi
P25	LPC exit pressure	0.5 psi
Ps3	Combustor static pressure	2.8 psi
P50	Engine exhaust pressure	0.3 psi
T2	Engine inlet temperature	1°F
T25	LPC outlet temperature	2.2°F
T30	HPC exit temperature	3°F
T50	Engine exhaust gas temperature	2.2°F

Construction of SVM and Kalman Filter

In our study, engine data (healthy and faulty) were generated by the C-MAPSS40K engine model. Therefore, C-MAPSS40K is used as the (surrogate) "engine" in Figure 2 for *both* engines. To implement the system depicted in Figure 2, the first step is to develop the SVM. To achieve this, the C-MAPSS40K model is linearized at 21 engine power settings and 5 Ram pressure ratios at Sea Level Standard Day (SLS). The state, input, output, and feedthrough matrices as well as the base points are saved as look-up tables and linearly interpolated for other operating conditions. The inputs and outputs of the SVM model are *corrected* engine parameters [13, 14].

The KF takes as input a set of measurement *residuals* between engine output parameters (i.e., N1, N2, Ps3, P50, T25, T50, and T30) and SVM estimates and produces a set of health parameters (or tuners) to minimize the residuals. The tuners are intended to represent engine health (interpreted as module efficiency and flow capacity changes) provided that the underlying model is accurate. Typical tuner selection usually includes a bias tuner to counter the inaccuracies in input parameters such as fuel flow. The choice of tuners is critical to the KF design and often depends on the available sensor measurement suite and parameter estimation requirements. Tuners are often selected through experiments or by following a more systematic optimal tuner selection process [15]. The tuner set listed in Table 2 has been chosen based on experience and testing results.

TABLE 2 Engine health parameters (tuners) for the KF

Tuner	Description
FanFC	Fan flow capacity modifier
LPCFC	LPC flow capacity modifier
HPCEff	HPC efficiency modifier
HPCFC	HPC flow capacity modifier
HPTEff	HPT efficiency modifier
WfBias	Fuel flow bias

Since the SVM is a direct linearization of the C-MAPSS40K model, its dynamic response and steady states are reasonably "close" to the engine data generated from C-MAPSS40k. This is not representative of real-world engine model fidelity given engine-to-engine variation that always exits. To make the simulation study more realistic, the following sensor biases have been added to the SVM in order to introduce a more severe modeling mismatch (Table 3).

With the modeling mismatch induced by sensor bias and linearization error, the tuner ($\Delta\eta^{engi}$ in Figure 2) values estimated by running the FOQA Flight #2 (Figure 3) are shown in Figure 4 below. It can be seen that the tuner values are corrupted by the modeling mismatch.

TABLE 3 Sensor biases applied to simulate modeling mismatch

Sensor	N1	N2	P25	Ps3	P50	T25	T30	T50
Bias	10	20	−0.3	2	0.4	−5	−20	15
Unit	RPM	RPM	psi	psi	psi	F	F	F

FIGURE 4 Estimated tuner values without engine calibration (Flight #2).

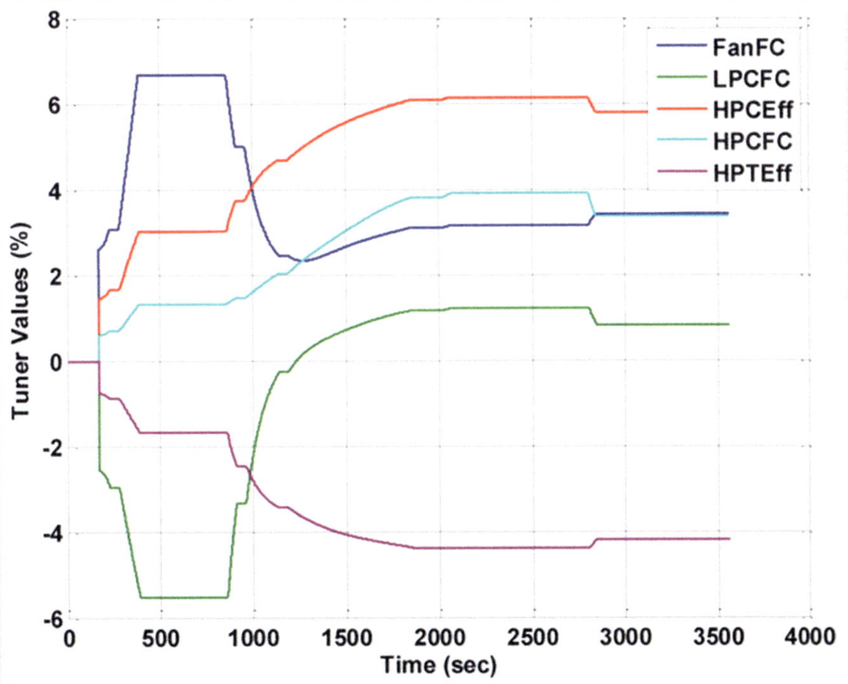

FIGURE 5 *QSS* status (Flight #2).

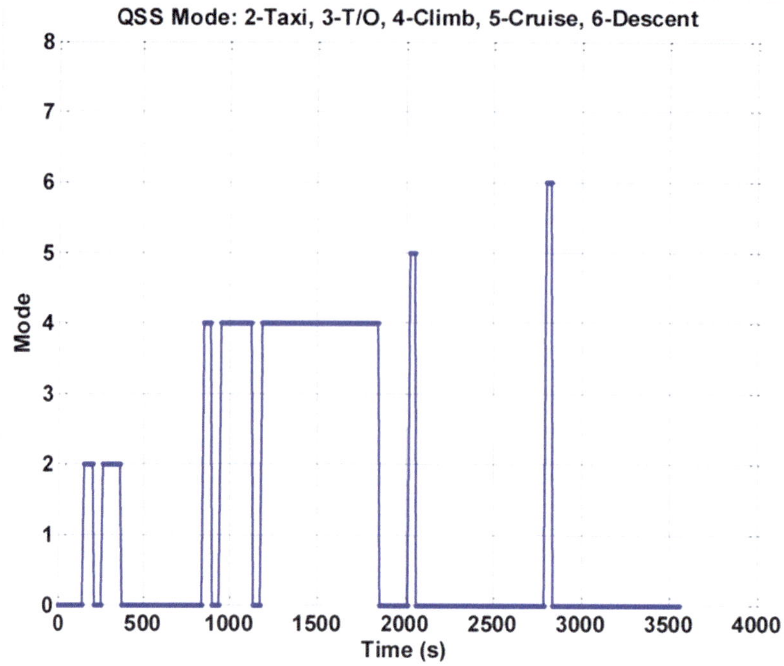

While the engine is healthy, the estimated HPT efficiency loss is at about 4% and High Pressure Compressor (HPC) efficiency has increased by 8%. These problems will be addressed by *QSS* Model Calibration in the following section.

The *QSS* status for this flight is shown in Figure 5. As shown in the figure, besides the very short cruise segments (at about 2000 s), a few other segments of flights from taxi, climb, and even descent have been selected based on the predefined *QSS* stability criteria. The stability criteria are defined by limiting the variance and ranges of certain gas path parameters such as N1, fuel flow, N2dot, etc. The data between these *QSS* segments are considered in transient, that is, outside the *QSS* Model Calibration region. The tuner values are held at last past values during transient operations.

The KF effect on measurement residuals is illustrated in Figure 6 using the EGT (T50) parameter as an example. The measured EGT is plotted in blue while the SVM output without KF is plotted in red. The difference between these two signals is essentially the residuals Δ^{eng} illustrated in Figure 2 (with the absence of Engine Calibration for now). When KF is engaged, the SVM (with KF in the loop, i.e., the lower SVM in Figure 2) EGT output is plotted in green and is very close to the measurement. Therefore, the SVM residuals are not as informative since the fault effect has been "absorbed" by the KF tuners.

Model Calibration

The particular model calibration scheme used for this system was an Artificial Neural Network (ANN)-based *empirical* modeling (EM) process (inspired by [16]), which captures the *difference* between the physics-based engine model (SVM) and the actual engine being monitored. Therefore, when the EM is incorporated into the engine model

FIGURE 6 EGT (T50) residuals (Flight #2).

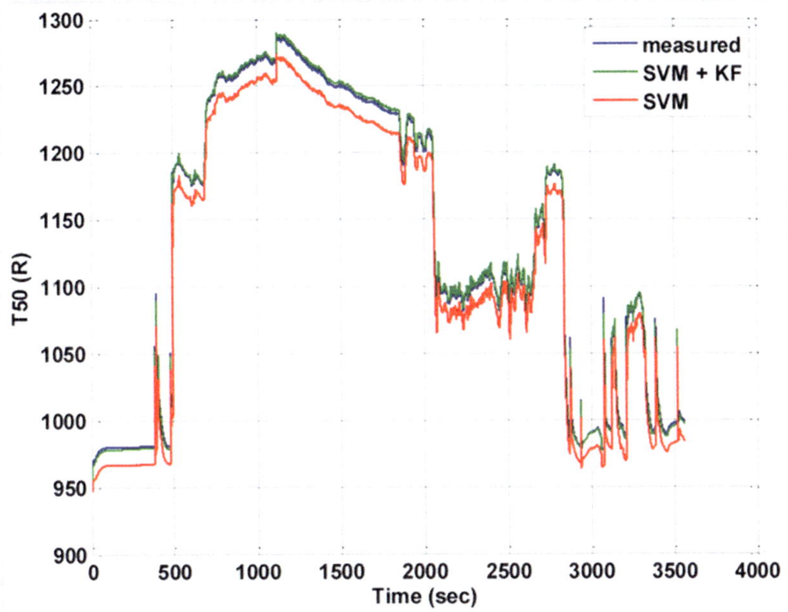

(indicated as Engine Calibration in Figure 2), the tuner corruption is mitigated, and the resulting KF tuners will reflect their intended function, that is, indicators of engine health.

For each engine, seven ANNs were constructed, one for each measured model output. The ANNs are three-layered feed-forward neural networks with five hidden neurons using a sigmoid activation function. Inputs to the ANNs are corrected N1, altitude, Mach number, and variable bleed valve (VBV) position. The ANN outputs are added to the outputs of SVM and are subtracted from engine measurements to generate the residuals for the KF. The ANN training is partitioned into multiple *QSS* segments in the flight envelope and uses a GMM-based training algorithm as described in [16]. In practice, an automated ANN training process needs to be incorporated either on board or off board (if full flight data can be recorded) to train the ANNs at installation, using a number of revenue service flights when the engines are relatively new (or have just come out of overhaul). Selected data points needs to be captured from each flight and aggregated for progressive training. In this study, we have implemented such an automated learning process based on the methods reported in [16] and applied to the engine data driven by FOQA inputs. From a total of 69 FOQA flights available, 35 were used to calibrate/train, and the remainder of flights (34) were used for validation.

After calibration was accomplished, the simulation using FOQA Flight #2 was run again with ANNs turned on and KF turned off. Figure 7 shows the comparison of the T50 (EGT) output from the SVM with ANNs calibration (green) and C-MAPSS40K engine data (blue) data. Since the modeling mismatch has been compensated by the ANNs, it can be anticipated that the KF tuners would be more confined and stay closer to zero. The KF tuner values are shown in Figure 8 in which all values are bounded by ±1%. Now the tuners are better grounded for engine health monitoring.

Three things are worth mentioning from this simulation results. First, FOQA Flight #2 was not included in the training data set. Therefore, the ANNs were not trained with

FIGURE 7 EGT (T50) residuals with engine calibration (Flight #2).

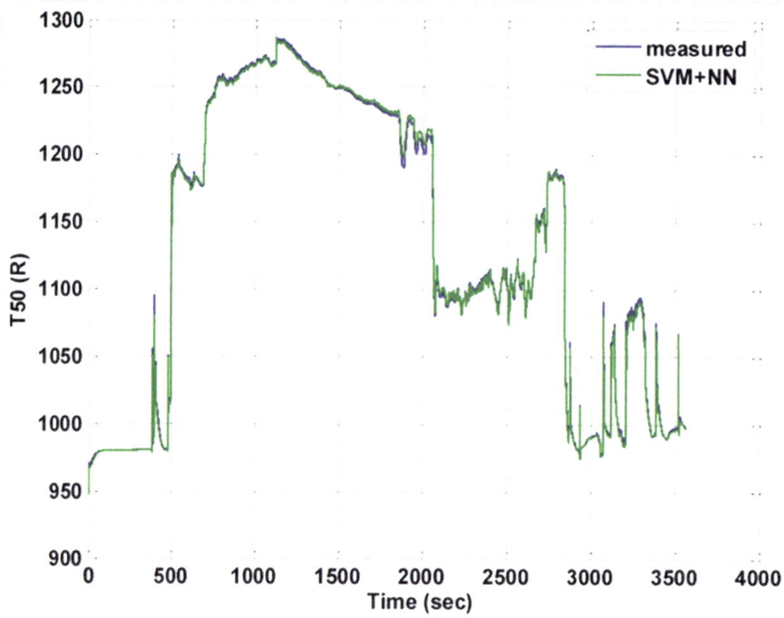

FIGURE 8 Estimated tuner values with engine calibration (Flight #2).

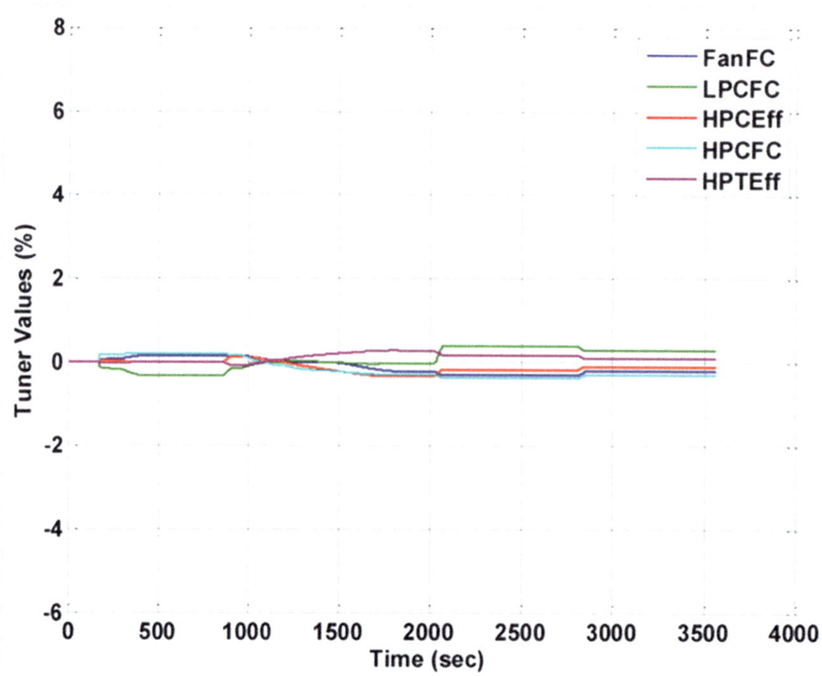

Flight #2 data. Secondly, although the SVM (with calibration) appeared to match the engine data well during transient as shown in Figure 7, this observation cannot be totally generalized. Training an empirical model to match engine dynamic response in all possible transient operations to a high degree of accuracy is still deemed infeasible. Fortunately, this is not a requirement for the FDI process under consideration. Last but not least, the remaining modeling error (after calibration) tends to cancel out when compared to the companion engine during a wide range of *QSS* operations due to the common modeling structure shared by both engines. This allows us to expand the *QSS* coverage to more flight regimes besides cruise, as well as to lower anomaly detection thresholds and false alarm rate.

Engine Nominal Deterioration

Engine calibration is performed during the first dozens of flights, after which the EM will be frozen and deployed to perform engine health monitoring. Over the life cycle of the engine, nominal deterioration takes place. Therefore, the originally matched model (SVM+ANN) will deviate gradually from actual engine performance over time. A mature EHM system should be robust to nominal deterioration. To evaluate the robustness of the cross-wing diagnosis approach, nominal deteriorations were introduced to both engines in the simulation study. The C-MAPSS simulation environment provides a set of empirical deterioration curves as shown in Figure 9 below.

FIGURE 9 Empirical engine health parameter deterioration curves (0, new; 50%, mid-life; 100%, end of life).

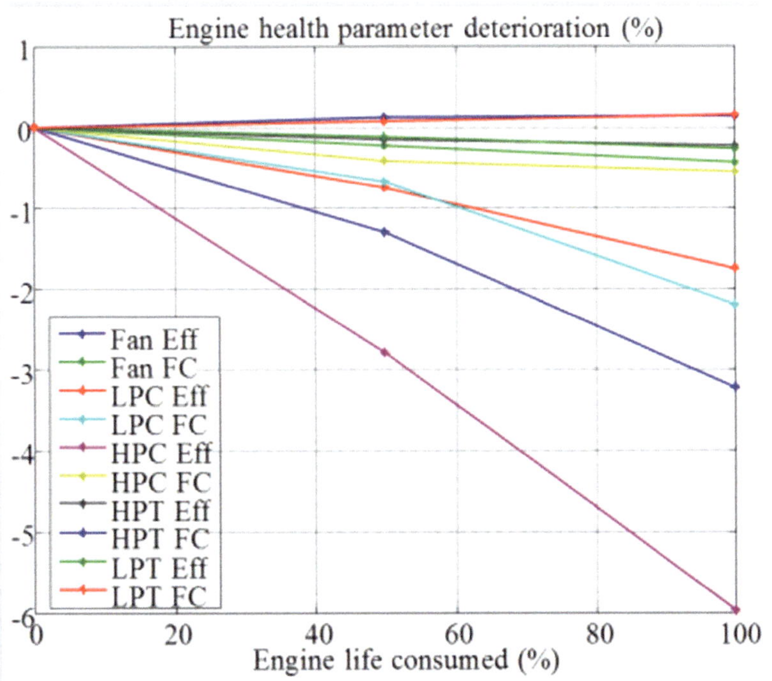

CHAPTER 12

FIGURE 10 Estimated tuner values at 50% life (Flight #2).

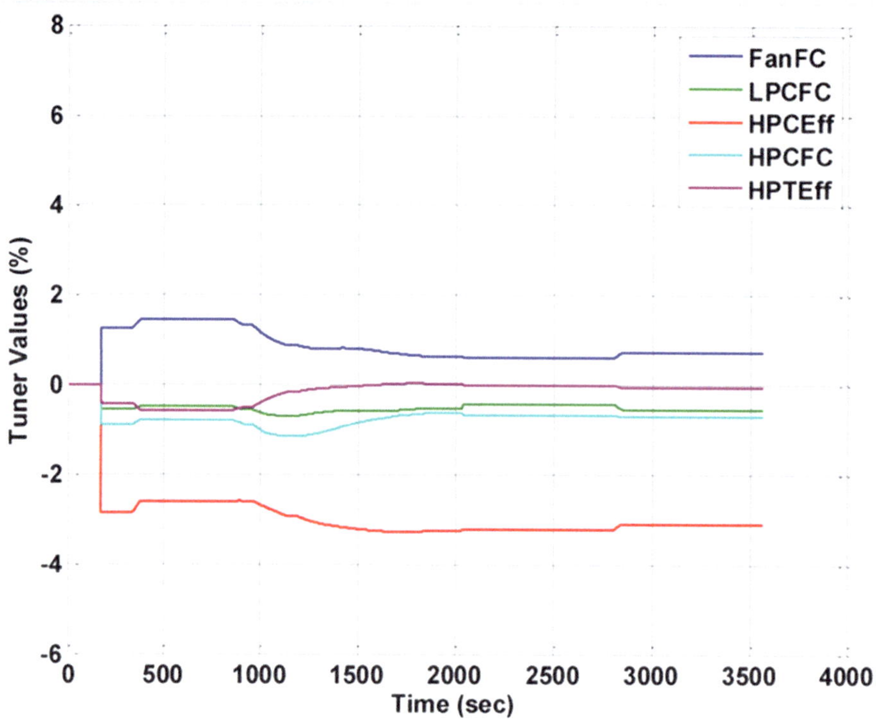

In our study, one of the engines is assumed at 50% life while the companion engine is assumed at 55%. Note that a fundamental assumption of the cross-wing diagnosis approach is that the two engines on the same aircrafts deteriorate at approximately *similar* rates. Figure 10 below shows the tuner values when 50% life consumption was introduced. The larger tuner deviations are results of performance deterioration.

Cross-Wing Fault Isolation

As mentioned earlier, fault isolation is based on the cross-wing parameter signature ($\Delta\Delta^{\times}$) described in Equation (2) and a procedure presented in [4]. To demonstrate the benefits of the cross-wing modeling error cancelation, a simple case study was shown in Figure A.2 in Appendix. A HPC fault (efficiency and flow capacity loss both at 3%) was induced at 500 s. In this figure, the top left and middle plots show the KF tuners from Engine 1 and Engine 2, respectively. The KF tuners of Engine 2 reflect the performance deterioration of a mid-life (55%) engine, while the KF tuners of Engine 1 reflect not only the nominal deterioration of a mid-life (50%) engine but also the fault effect induced at 500 s for 3% HPC performance loss. The cross-wing tuners are shown in the upper right plot. The cancelation of nominal deterioration effect is apparent. The cross-wing tuners for HPC efficiency and HPC flow capacity are about −3% which agrees with the fault simulated. However, this is only because both HPC efficiency and flow capacity are included in our tuner selection. For all other faults, all the tuners will be affected due to a well-known smearing effect. Thus, the cross-wing tuners alone are not effective for fault isolation. The bottom plots show the Engine

1 SVM residuals, Engine 2 SVM residuals, and cross-wing ΔΔ residuals, respectively. The cross-wing ΔΔ residuals provide the fault signature for isolation. The final diagnosis result is a result of majority voting using all *QSS* data points at each time step aggregated at the end of the flight. In this case, the fault was correctly isolated with good confidence (90.1%).

Cross-Wing Diagnosis Monte Carlo Simulations

A comprehensive Monte Carlo Simulation (MCS) was conducted to provide verification for the cross-wing approach. Eighteen fault modes were simulated as listed in Table 4 below. The coupling factor (CF) for module fault is defined as the ratio between flow capacity and efficiency, that is, CF = flow capacity/efficiency. Note that VBV position is in fraction, that is, 1 means 100% stroke length. For CMAPSS40K, VBV is physically closed below 15.4% stroke.

In the MCS, 100 randomly generated instances for *each* fault appearing in Table 4 (with random component fault coupling) were introduced on a mid-life 50% Engine 1 and compared to (non-faulted) Engine 2 at a mid-life 55%. For all simulations, the fault was induced at 500 s typically during climb. Sensor output data contained randomly generated noise at levels indicated in Table 1. The aforementioned sensor biases as shown in Table 3 were included. A total of *10 flight profiles* in the FOQA data set were randomly chosen for the simulation study, and the 100 random faults were uniformly distributed over the 10 flight profiles. It should be noted that only one out of the ten flight profiles was included in the ANN training set while the remaining nine were all from the validation data set. The actual flight profiles are provided in Figure A.1 in the Appendix. The resultant confusion matrix appears in Table 5 and also in the Appendix.

TABLE 4 Random parameter ranges for the Monte Carlo simulation

Class	Fault mode	Random fault size range (uniform dist.)
1	Fan fault	Eff. loss: [1~5]%, CF: [1.1~1.5]
2	LPC fault	Eff. loss: [1~5]%, CF: [1.1~1.5]
3	HPC fault	Eff. loss: [1~5]%, CF [0.8~1.2]
4	HPT fault	Eff. loss: [1~5]%, CF [−0.9~−0.5]
5	LPT fault	Eff. loss: [1~5]%, CF [−0.8~−0.4]
6	VBV actuator bias	[0.2~0.4]
7	VSV actuator bias	−[1.5~6] degrees
8	Fuel flow bias	±[0.08~3] lb/s
9	N1 sensor bias	±[10~90] RPM
10	N2 sensor bias	±[15~100] RPM
11	T2 sensor bias	±[3~10] F
12	P2 sensor bias	±[0.6, 2] psi
13	T25 sensor bias	±[6, 20] F
14	P25 sensor bias	±[1.6, 5] psi
15	T30 sensor bias	±[6, 30] F
16	Ps3 sensor bias	±[4, 15] psi
17	T50 sensor bias	±[8, 60] F
18	P50 sensor bias	±[0.5, 4] psi

TABLE 5 Diagnostic confusion matrix for Monte Carlo simulation (Engine 1, 50% life; Engine 2, 55% life)

Classified \ Truth	Fan	LPC	HPC	HPT	LPT	VBV	VSV	Wf	N1	N2	T2	P2	T25	P25	T30	Ps3	T50	P50	No fault
Fan	100	0	0	0	0	0	0	0	0	0	0	0	0	0	0	0	0	0	0
LPC	0	98	0	0	0	0	0	0	0	0	0	0	0	0	0	0	0	0	0
HPC	0	0	99	0	0	0	0	0	0	0	0	0	0	0	0	0	0	0	0
HPT	0	0	0	100	0	0	0	0	0	0	0	0	0	0	0	0	0	0	0
LPT	0	0	0	0	100	0	0	0	0	0	0	0	0	0	0	0	0	0	0
VBV	0	2	0	0	0	100	0	0	0	5	0	0	0	0	0	0	0	0	0
VSV	0	0	0	0	0	0	78	0	0	3	0	0	0	0	0	0	0	0	0
Wf	0	0	0	0	0	0	0	100	0	0	0	0	0	0	0	0	0	0	0
N1	0	0	0	0	0	0	0	0	95	0	0	0	0	0	0	0	0	0	0
N2	0	0	0	0	0	0	2	0	0	62	0	0	0	0	0	0	0	0	0
T2	0	0	0	0	0	0	0	0	0	0	100	0	0	0	0	0	0	0	0
P2	0	0	0	0	0	0	0	0	5	1	0	100	0	0	0	0	0	0	0
T25	0	0	0	0	0	0	0	0	0	0	0	0	87	0	0	0	0	0	0
P25	0	0	1	0	0	0	20	0	0	0	0	0	0	100	0	0	0	0	0
T30	0	0	0	0	0	0	0	0	0	0	0	0	0	0	99	0	0	0	0
Ps3	0	0	0	0	0	0	0	0	0	3	0	0	0	0	0	100	0	0	0
T50	0	0	0	0	0	0	0	0	0	2	0	0	0	0	1	0	93	0	0
P50	0	0	0	0	0	0	0	0	0	0	0	0	0	0	0	0	0	100	0
No fault	0	0	0	0	0	0	0	0	0	18	0	0	13	0	0	0	7	0	100

Based on the numbers in the confusion matrix, the following overall diagnostic performance (for the range of faults simulated) can be calculated (Table 6).

TABLE 6 Cross-wing diagnostic performance

Misdetection rate	2.1%
False alarm rate	0%
Isolation accuracy (faults)	95%

Summary/Conclusions

We have provided a broad discussion on the relative benefits and technical challenges pertaining to the monitoring and analysis of full flight data. The primary benefit of reducing time latency for FDI to allow maintenance advisory by end of flight reduces the possibility for collateral and secondary damage, thereby supporting the basic drivers for engine condition monitoring of reducing IFSDs and UERs and extending time on wing. To enable this endeavor, we have introduced a methodology for cross-wing comparison and analysis. The use of companion engine data provides a worthy example of the power of *information fusion* for engine diagnostics and FDI. It was noted that accurate engine modeling is a pivotal element and primary technical challenge to achieving successful diagnostic results, especially for full flight data analyses containing both steady-state and transient maneuvers. The effective cross-wing cancelation of errors, introduced in part by model infidelity, permits a more robust and timely diagnosis of in-flight faults than what can be achieved currently with legacy methods.

References

1. Volponi, A., "Gas Turbine Engine Health Management: Past, Present and Future Trends," *Journal of Engineering for Gas Turbines & Power* 136, no. 5, May 2014.
2. Urban, L. and Volponi, A., "Mathematical Methods of Relative Engine Performance Diagnostics," SAE Technical Paper 922048, 1992, doi:10.4271/922048.
3. Doel, D.L., "TEMPER - A Gas Path Analysis Tool for Commercial Jet Engines," *Proceedings of the American Society of Mechanical Engineers Turbine and Aero-Engine Congress*, Cologne, Germany, June 1-4, 1992, ASME Paper 92-GT-315.
4. Volponi, A., "Foundations of Gas Path Analysis I-II," Extending Gas Path Analysis Coverage for Other Fault Conditions, Von Karman Institute for Fluid Dynamics Lecture Notes 2003-01, January 2003.
5. Volponi, A. and Rajamani, R., "Hybrid Models for Engine Health Management," *Machine Learning and Knowledge Discovery for Engineering Systems Health Management* (CRC Press, 2012), Chapter 12.
6. Borguet, S., Dewallef, P., and Leonard, O., "A Way to Deal with Model-Plant Mismatch for Reliable Diagnosis in Transient Operation," *Proceedings of the American Society of Mechanical Engineers Turbine and Aero-Engine Congress*, Barcelona, Spain, May 8-11, 2006, Paper No. GT2006-90412.
7. Armstrong, J. and Simon, D.L., "Constructing an Efficient Self-Tuning Aircraft Engine Model for Control and Health Management Applications," NASA/TM-2012-217806, December 2012.
8. Panov, V., "Auto-Tuning of Real-Time Dynamic Gas Turbine Models," *Proceedings of the ASME Turbo Expo 2014 Technical Conference*, June 16-20, 2014, Paper No. GT2014-25606; Volponi, A., "Use of Hybrid Engine Modeling for On-Board Module Performance Tracking," *IGTI Turbo Expo 2005*, Reno, NV, June 2005, ASME GT2005-68169.

9. May, R.D., Csank, J., Lavelle, T.M., Litt, J.S. et al., "A High-Fidelity Simulation of a Generic Commercial Aircraft Engine and Controller," *Proceedings of 46th AIAA/ASME/SAE/ASEE Joint Propulsion Conference and Exhibit*, July 25-28, 2010, Paper No. AIAA-2010-6630.
10. Liu, Y. et al., "User's Guide for the Commercial Modular Aero-Propulsion System Simulation (C-MAPSS)," NASA/TM-2012-217432, March 2012.
11. Simon, D.L., "Propulsion Diagnostic Method Evaluation Strategy (ProDiMES) User's Guide," NASA/TM 2010-215840, January 2010.
12. US DOT: Advisory Circular, "Flight Operations Quality Assurance (FOQA)," AC No.: 120-82, April 12, 2004.
13. Volponi, A.J., "Gas Turbine Parameter Corrections," *Transactions of the ASME, Journal of Engineering for Gas Turbines and Power* 121, no 4 (October 1999): 613-621.
14. Kurzke, J., "Model Based Gas Turbine Parameter Corrections," *Proceedings of 2003 ASME TURBO EXPO: Power for Land, Sea, & Air*, Atlanta, GA, June 16-19, 2003, Paper GT2003-38234.
15. Simon, D.L. and Garg, S., "Optimal Tuner Selection for Kalman Filter-Based Aircraft Engine Performance Estimation," *Journal of Engineering for Gas Turbines and Power* 132 (March 2010).
16. Volponi, A.J. et al., "Empirical Tuning of an On-Board Gas Turbine Engine Model for Real-Time Module Performance Estimation," *Transactions of the ASME, Journal of Engineering for Gas Turbines and Power* 130, no. 2 (February 2008).

Contact Information

Allan J. Volponi, Senior Fellow, Pratt & Whitney
400 Main St., M/S 162-15, East Hartford, CT 06108
allan.volponi@pw.utc.com
Tele: 860-557-2421

Liang Tang, Principal Engineer, Pratt & Whitney
400 Main St., M/S 162-15, East Hartford, CT 06108
liang.tang@pw.utc.com
Tele: 860-565-5348

Definitions/Abbreviations

ANN - Artificial Neural Network
CBM - Conditioned-Based Maintenance
CF - Coupling Factor
C-MAPSS - Commercial Modular Aero-Propulsion System Simulation
DOD - Domestic Object Damage
Eff - Efficiency
EGT - Exhaust Gas Temperature
EHM - Engine Health Monitoring
EM - Empirical Model
FC - Flow Capacity
FDI - Fault Detection and Isolation

FOD - Foreign Object Damage
FOQA - Flight Operations Quality Assurance
GMM - Gaussian Mixed Model
GPA - Gas Path Analysis
HPC - High Pressure Compressor
HPT - High Pressure Turbine
IFSD - In-Flight Shutdown
KF - Kalman Filter
LPC - Low Pressure Compressor
LPT - Low Pressure Turbine
MCS - Monte Carlo Simulation
N1 - Fan Speed
N2 - Core Speed
PLA - Power Lever Angle
P2 - Fan Inlet Pressure
P25 - LPC Exit Pressure
Ps3 - Combustor Static Pressure
P50 - Engine Exhaust Pressure
QSS - Quasi-Steady State
SVM - State Variable Model
T2 - Engine Inlet Temperature
T25 - LPC Outlet Temperature
T30 - HPC Exit Temperature
T50 - Engine Exhaust Gas Temperature
UER - Unscheduled Engine Removal
VBV - Variable Bleed Valve
VSV - Variable Stator Vane
Wf - Fuel Flow

Appendix

FIGURE A.1 FOQA flight profiles used in Monte Carlo simulations.

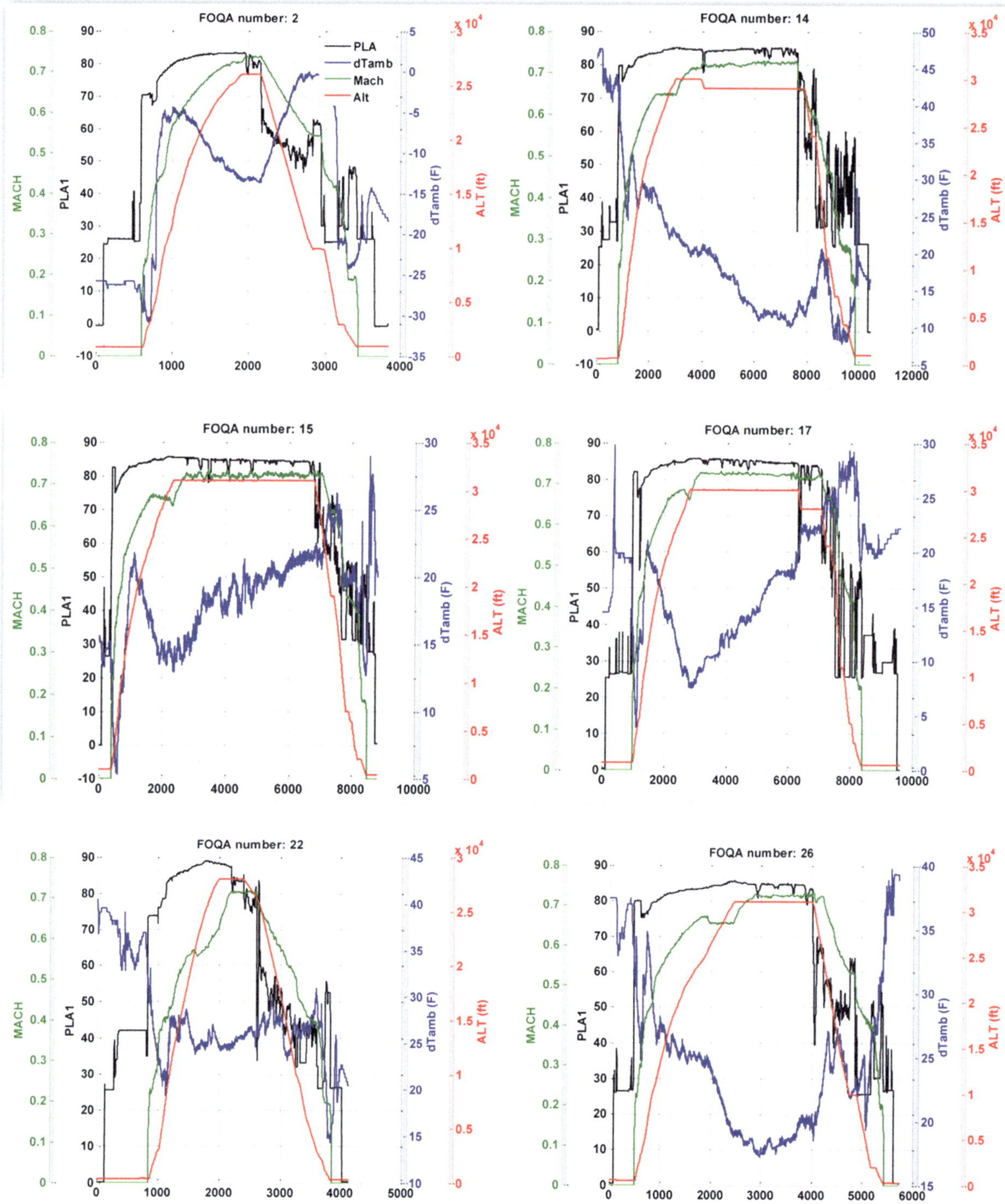

FIGURE A.1 *(Continued)* FOQA flight profiles used in Monte Carlo simulations.

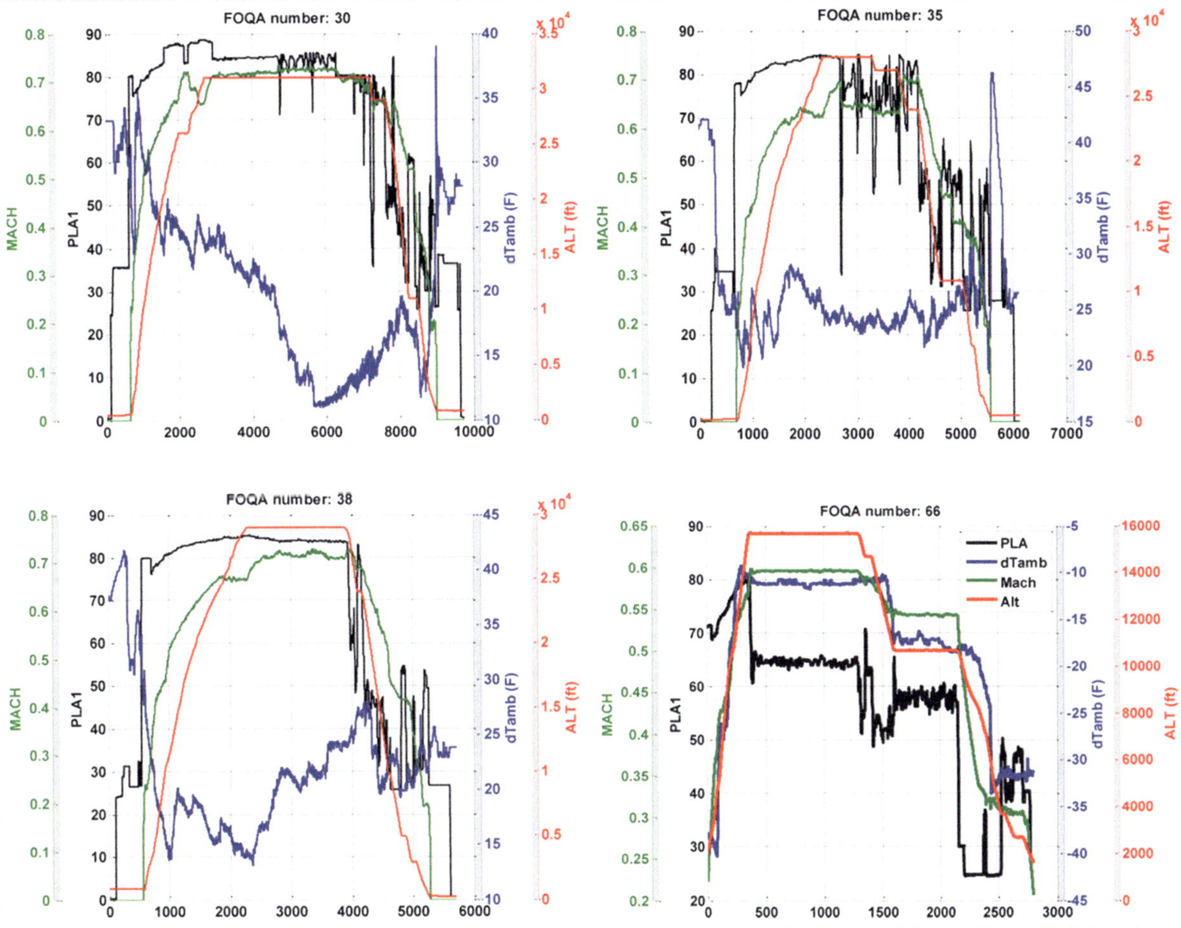

FIGURE A.2 Cross-wing diagnosis result for a HPC fault.

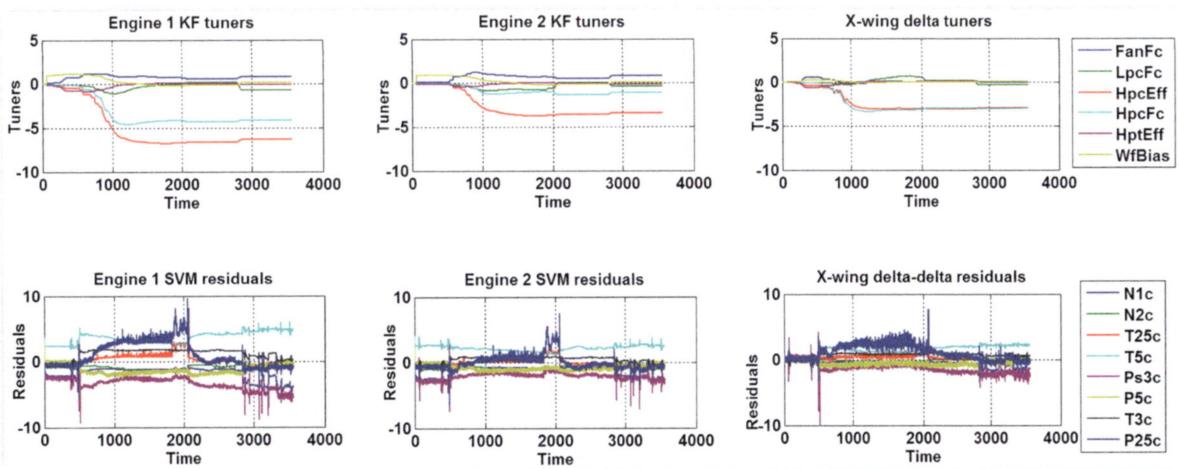

CHAPTER 13

Establishment and Application of Aeroengine Lubricating Oil System Model Based on Flight Test Data

Mingming Ma
CFTE

A lubricating system modeling method based on flight test data is proposed in this paper. Artificial neural network (ANN) model based on a large number of flight test data is trained and validated, and models of six lubricating system parameters in all engine operation settings and whole flight envelope are established. Model results are in good agreement with flight test results, which shows feasibility and effectiveness of the presented modeling method. The model results are packaged in dynamic link library (DLL), and the coordination between calculating model and GDAS is accomplished. Comparison of model and flight test results in real-time monitoring of flight test comes true; thus online trend monitoring of oil parameters is implemented and applied. Additionally, input parameters are gradually decreased as new input parameter group of ANN structure. Oil parameter model is trained and validated again with the new group of parameters, until leading to unacceptable bias between model and flight test results. The main influencing factor is the combination of the least input parameters and its maximum difference between model and flight test approximates original modeling results. Therefore, an approach to determine the main influencing parameters of aeroengine lubricating system is developed.

Introduction

Lubricating system is an important subsystem of an aeroengine, and it affects the operation reliability and safety of an engine. Lubricating system is required to provide lubricating and cooling, reduce friction and corrosion, and remove heat of the high-speed rotating components, and thus ensure its long duration operation in normal condition [1, 2]. The essential factor for flight evaluating, condition monitoring, and maintenance of lubricating system is to determine its operation status on different flight conditions and engine operation settings [3, 4]. Lubricating system modeling is an effective approach to investigate its operation status.

There are three main methods to be utilized on lubricating system modeling.

The first is to establish mathematical models of components and subsystems by analyzing lubricating system structure and component characteristics (e.g., oil pump supplying characteristic, oil tube resistance characteristic, heat exchanger exchanging characteristic, oil-gas separator pressure loss characteristic, and so on). Then, they will be combined to process iterative calculation and finally integrated into lubricating system model [5, 6, 7].

The second is to simulate by Computational Fluid Dynamics (CFD). Lubricating system is divided into an element-and-node network according to structure and chamber distribution. The inner flow within the network will be simulated with limited elements as well as medium types, and steady-state operation parameters of lubricating system will be resolved with given boundary condition. Consequently, the steady-state lubricating system is attained [8, 9].

The last one is to utilize data mining methods based on lubricating system data, such as artificial neural network (ANN) [10, 11, 12], support vector machines (SVM) [13, 14, 15], and so on. An aeroengine lubricating oil metal element concentration prediction method based on the Elman process ANN is proposed [12]. A novel aeroengine lubricating metal content forecasting method based on SVM is presented [15].

A lubricating system modeling method based on flight test data is proposed in this paper. A large number of flight test data is trained and validated by applying ANN, and models of six lubricating system parameters in all engine operation settings and whole flight envelope are established, including oil supply pressure, oil differential pressure, scavenge oil temperature of rear bearing cavity, etc. Model results are in good agreement with flight test results, which shows feasibility and effectiveness of the presented modeling method. The model results are packaged in dynamic link library (DLL), and the coordination between calculating model and Ground Data Acquisition System (GDAS) is accomplished. Comparison of model and flight test results in real-time monitoring of flight test comes true, and thus online trend monitoring of oil parameters has been realized and applied. In addition, input parameters are gradually decreased as new input parameter group of ANN structure. Sample data is trained and validated again with the new group of parameters, until leading to unacceptable bias between model and flight test results. Ultimately, an approach to determine the main influencing parameters of aeroengine lubricating system is developed by utilizing the presented modeling method.

Lubricating System Description and Modeling Method Instruction

Lubricating System Description

The lubricating system in the study is a self-contained re-circulatory one and arranged on a low-bypass-ratio, twin-spool, afterburning turbo fan engine. It consists of a supply

FIGURE 1 Simplified scheme and measurement locations of a lubricating system.

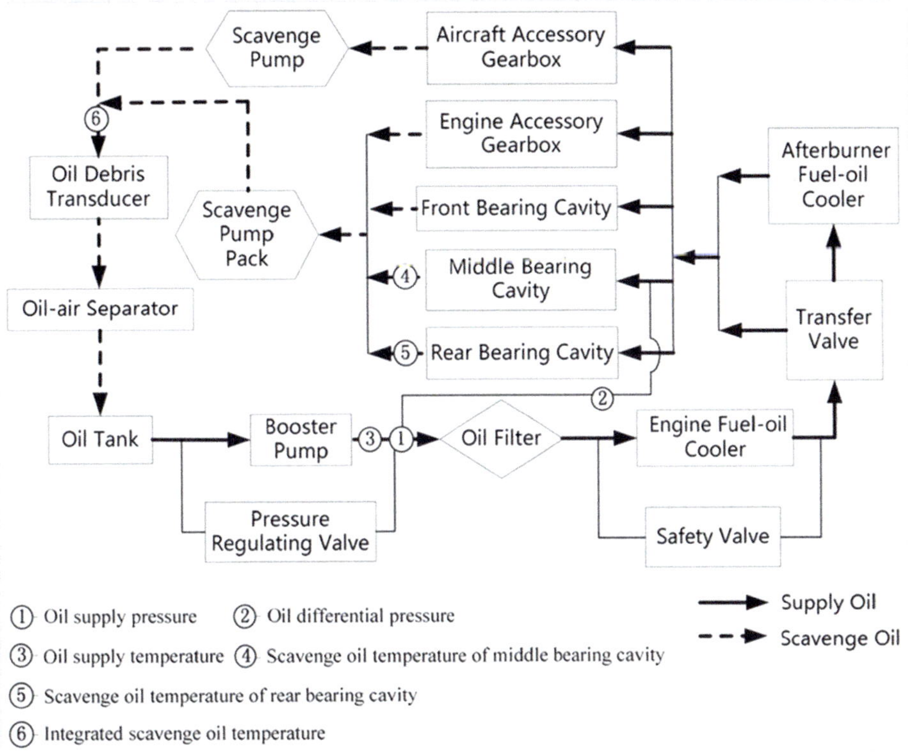

subsystem, a scavenge subsystem, a vent subsystem, and a pressurized sealing subsystem, shown as Figure 1.

The lubricating oil in oil tank is divided into three paths after pressurized by a booster pump, filtered by oil filter, and heat dissipated by engine fuel-oil cooler, which are respectively supplied to three bearing cavities of the engine, engine accessory gearbox, and aircraft accessory gearbox. When afterburner is working on, in addition, the oil will also be cooled by afterburner fuel-oil cooler. Three paths scavenge oil is pressurized by scavenge pump pack and is back to the oil tank after separation of oil and air by oil-air separator. An oil debris transducer is used to detect the metal content in the scavenge oil.

The primary purpose of lubricating system is to lubricate and cool the engine's main shaft bearings (five bearings located in front, middle, and rear bearing cavity, respectively), sealing equipment, central gearbox, engine accessory gearbox, and aircraft accessory gearbox, to ensure their satisfactory operation.

As presented in Figures 1 and 6 parameters have been measured during flight test of an aeroengine, which are oil supply pressure P_{os}, oil differential pressure P_{od}, represented the pressure difference between oil supply pressure and middle bearing cavity pressure, oil supply temperature T_{os}, scavenge oil temperature of middle bearing cavity T_{om}, scavenge oil temperature of rear bearing cavity T_{or} and integrated scavenge oil temperature T_{oi}. The operation status of aeroengine lubricating system could be represented by these parameters during flight test.

ANN Modeling Method Instruction

ANN has been trained to perform complex functions in various fields of application including parameter identification and forecasting and pattern recognition. ANN has strong nonlinear reflection ability, as well as self-adaption and self-learning ability on complex uncertain problems, and is highly fault tolerant, which is especially suitable for resolving the problem that there are a large number of data samples and the output-input relationship is not clear.

It is demonstrated that any nonlinear system can be approximated by selecting and training hyperbolic tangent transfer function on the hidden layer and pure linear transfer function on the output layer of three-layer ANN. Additionally, ANN training result is affected by the selection of neuron number. According to experience, if the neuron number is "n" on the previous layer, there will be "2n+1" neuron number on the current layer. Three-layer ANN structure is employed to develop lubricating system model in this paper.

Flight Test Data Instruction

The data samples are obtained from a flight test of an aeroengine, including cold and hot climatic condition test, etc. The total sample size S_t is 27,648, representing 27,648 different points of flight conditions and engine operating settings. The points almost cover the whole flight envelop and the engine operation settings, involving various steady states, airstart, accelerating, decelerating, and afterburner on and off. Due to data samples of the negative acceleration condition are not enough to support the application of ANN, so oil parameter modeling on this topic is not studied in this paper. As shown in Table 1, in addition to the six oil parameters (P_{os}, P_{od}, T_{os}, etc.), parameters of flight condition, flight attitude, and engine operation settings should be measured, for example, flight altitude H, flight Mach number Ma, total air temperature (TAT), roll angle ϕ, longitudinal acceleration Nx, lateral acceleration Ny, normal acceleration Nz, power lever angle (PLA), engine high-pressure compressor speed n_2, combustion fuel pressure P_f, high-pressure compressor discharge pressure P_3, and so on. All the measurements have been examined before modeling.

TABLE 1 Minimum and maximum values of measurements

Parameters	Minimum value	Maximum value
Flight altitude	288 m	17,528 m
Flight Mach number	0.19	1.79
Total air temperature	−53.6°C	82.5°C
Roll angle	−179.4°	179.9°
Longitudinal acceleration	−0.85 g	1.08 g
Lateral acceleration	−0.67 g	0.49 g
Normal acceleration	−0.30 g	8.47 g
Power lever angle	0°	114.8°
High-pressure compressor speed	10.0%	103.7%
Combustion fuel pressure	78.5 kPa	6731.0 kPa
High-pressure compressor discharge pressure	62.0 kPa	3513.6 kPa
Oil supply pressure	70.8 kPa	608.3 kPa
Oil differential pressure	3.1 kPa	306.8 kPa
Oil supply temperature	56.6°C	143.5°C
Scavenge oil temperature of middle bearing cavity	107.4°C	176.0°C
Scavenge oil temperature of rear bearing cavity	118.3°C	196.5°C
Integrated scavenge oil temperature	108.7°C	165.9°C

Modeling of Oil Parameters

Analysis and Determination of Main Influencing Parameters on Lubricating Parameters

Lubricating system parameters are closely related to fight conditions, accelerations, and engine operation settings. Before utilizing ANN, influencing factors on aeroengine lubricating parameters must be confirmed. In this paper, lubricating system working principle and characteristics will be analyzed to confirm influencing parameters, which are input parameters of the lubricating system modeling.

With the changing of flight conditions, engine operating parameters, especially high-pressure compressor speed, will vary. The higher the compressor speed is, the more the heat will be produced by friction, and the faster the oil circulation will be required; as a result, the higher the oil pressure will be possibly achieved. When the flight accelerations change, the relative location of oil tank and oil nozzle will be changed consequently, as well as the oil supply quantity and pressure. However, oil booster pump is directly driven by high-pressure compressor. Therefore, the oil supply quantity or pressure is mainly affected by engine settings instead of flight conditions. In conclusion, inputs of P_{os} and P_{od} modeling involve eight parameters, which are *H*, *Ma*, *TAT*, *Nx*, *Ny*, *Nz*, *PLA*, and n_2. The purpose of choosing *PLA* as an input is to distinguish engine steady state and transient state. Structure of oil supply pressure P_{os} ANN model is shown in Figure 2.

As mentioned previously, the changing of flight conditions and engine operating settings will lead to variations of oil supply quantity, which will affect the oil temperature. Furthermore, fuel parameters, such as fuel temperature and fuel flow mass, will also affect oil temperature because of the heat exchanging between fuel and oil. To summarize, inputs of T_{os}, T_{om}, T_{or} and T_{oi} modeling separately involve eight parameters, which are *H*, *Ma*, *TAT*, ϕ, P_{od}, *PLA*, n_2, and fuel-air pressure difference between combustion fuel pressure and high-pressure compressor discharge pressure $P_{f\text{-}3}$. Fuel-air pressure difference is represented fuel flow mass according to the fuel nozzle

FIGURE 2 Structure of oil supply pressure ANN model.

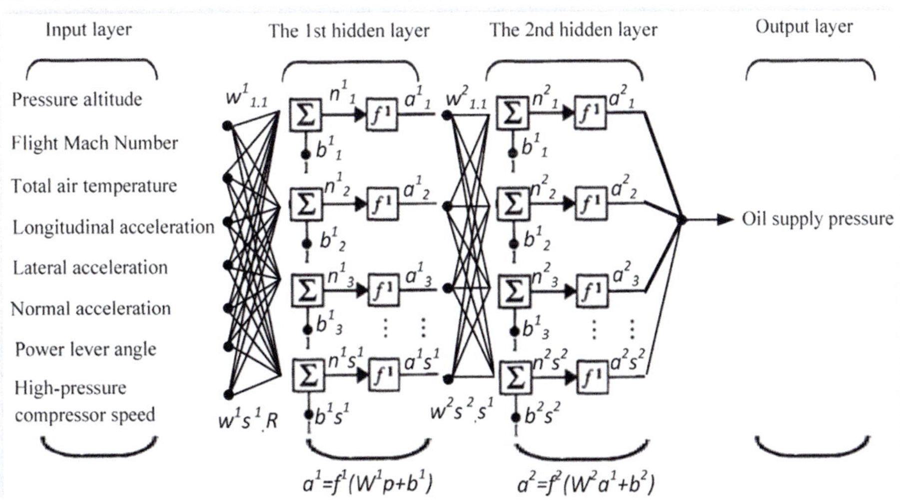

characteristics. In this paper, the impact of fuel temperature on oil scavenge temperature is ignored because of lack of data.

Model Results of Oil Parameters

Based on the analysis and determination of influencing factors of oil parameters as inputs, oil parameter model is trained and validated; eventually, models of six oil parameters are obtained. Distributions of difference value between model and flight test results of six parameters in the study are illustrated from Figures 3, 4, 5, 6, 7, and 8, respectively, and their difference ranges are given in Table 2.

In Figures 3, 4, 5, 6, 7, and 8, the ordinate r refers to ratio of distribution, defined as distribution size S_d to total sample size S_t of flight test data.

It indicates that model results are in good agreement with flight test results, which shows the feasibility and effectiveness of the presented modeling method.

FIGURE 3 Distribution of difference values between model and flight test results of P_{os}.

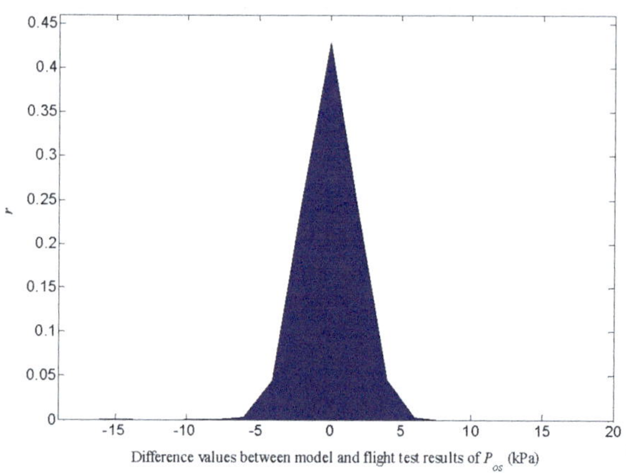

FIGURE 4 Distribution of difference values between model and flight test results of P_{od}.

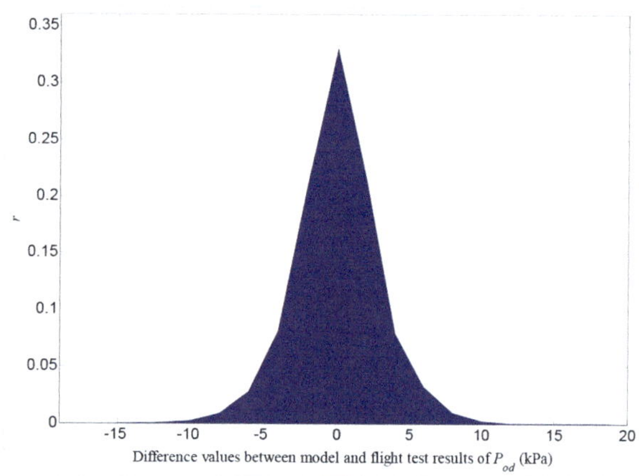

FIGURE 5 Distribution of difference values between model and flight test results of T_{os}.

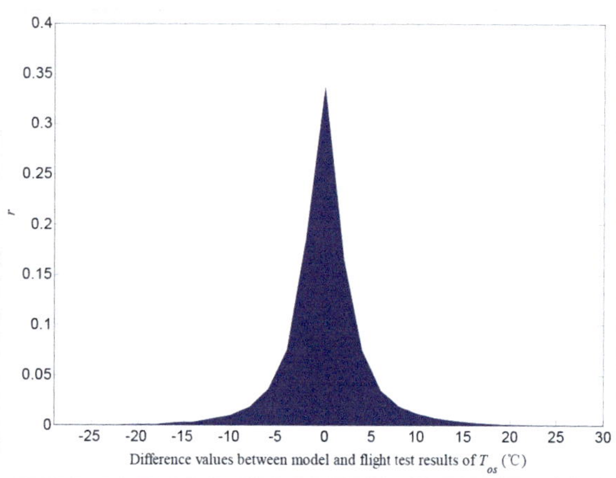

FIGURE 6 Distribution of difference values between model and flight test results of T_{om}.

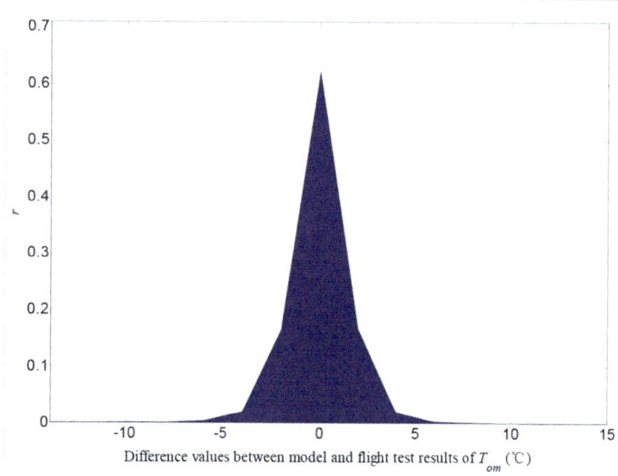

FIGURE 7 Distribution of difference values between model and flight test results of T_{or}.

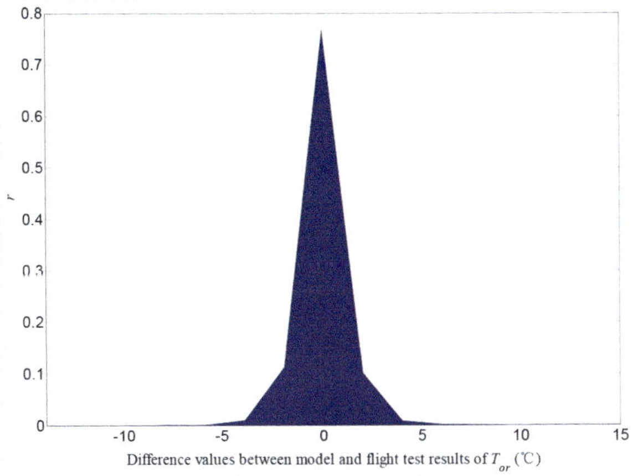

FIGURE 8 Distribution of differences value between model and flight test results of T_{oi}.

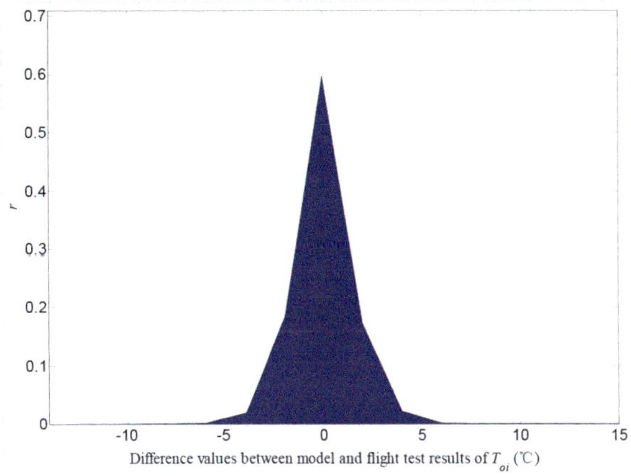

TABLE 2 Difference ranges between model and flight test results of oil parameters

Parameters	Difference range between model and flight test results	
	Minimum value	Maximum value
Oil supply pressure	−17.7 kPa	16.2 kPa
Oil differential pressure	−14.2 kPa	18.9 kPa
Oil supply temperature	−24.0°C	28.8°C
Scavenge oil temperature of middle bearing cavity	−11.8°C	9.2°C
Scavenge oil temperature of rear bearing cavity	−7.2°C	10.3°C
Integrated scavenge oil temperature	−8.1°C	12.6°C

Application of Oil Parameter Model

Online Trend Monitoring of Oil Parameters

Currently, limit-value monitoring is widely used on lubricating system condition monitoring, by comparing oil parameters, such as oil differential pressure and temperature, with their corresponding limit values, to estimate whether the engine operates satisfactorily or not. However, when the trend of oil parameter evolution becomes abnormal within limit settings, it will not be distinguished by limit-value monitoring and may possibly lead to serious fault or even failure if it happens during flight test.

An important means to resolve this problem is online trend monitoring, which compares lubricating system test parameter with its baseline on real time. When the difference value appears suddenly or becomes unacceptable, the oil system is supposed to be unhealthy [16, 17]. In contrast with limit-value monitoring, trend monitoring takes engine parameter baseline and individual discrepancy into consideration [18, 19, 20]. Trend monitoring is able to implement malfunction forecasting and, therefore, enhance reliability and safety supportability of flight test [21, 22].

The oil parameter model, structured previously, is packaged in DLL. The DLL and corresponding monitoring screen are applied to the GDAS.

FIGURE 9 Variation of relevant parameters with time in an entire flight.

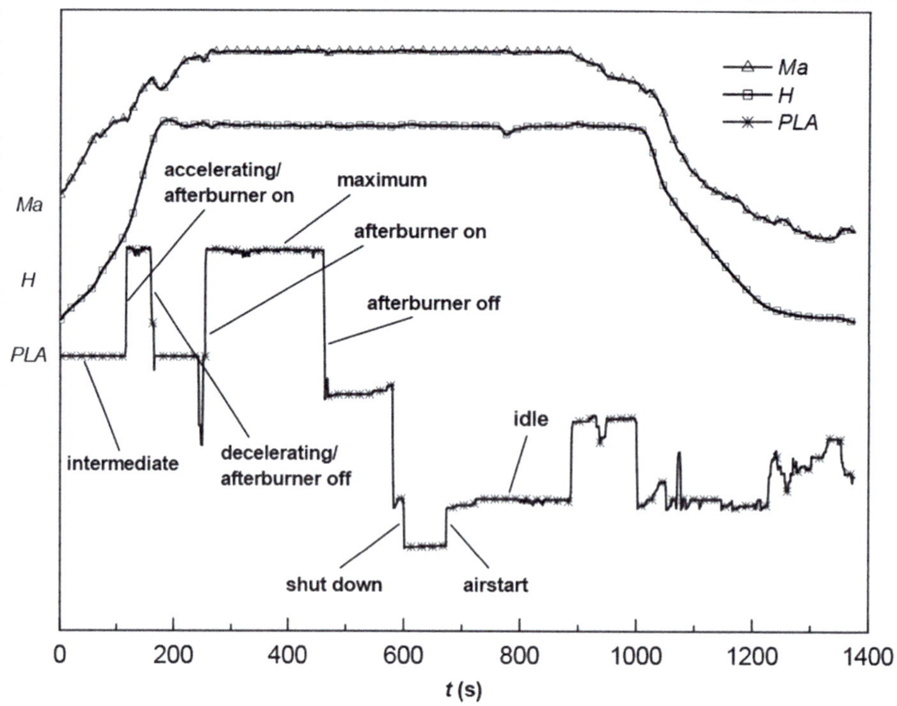

During flight test, DLL is invoked by GDAS to calculate measurements on real time. Once DLL calculation has been completed and return values have been obtained, their trend will be compared with flight test results and displayed on the monitoring screen.

Figure 9 shows the variation of relevant parameters with time in an entire flight. From the value of *PLA*, the engine settings can be read, which include steady state (idle, intermediate, and maximum settings), accelerating, decelerating, afterburner on and off, shutdown and airstart, etc. Comparisons between model and flight test results of oil differential pressure P_{od} and scavenge oil temperature of rear bearing cavity T_{or} are presented in Figures 10 and 11, respectively, which are well agreed. Therefore, accuracy of oil parameter model as well as feasibility and practicability of its application in online trend monitoring are further demonstrated.

Determination of Main Influencing Factors on Oil Parameters

Oil parameter model established in this paper is considered as baseline results, and its input parameter is defined as baseline input parameters. Then, input parameters are gradually decreased as new input parameter group of ANN structure. Oil parameter model is trained and validated again with the new group of parameters, and the results would be compared with baseline results. The maximum difference between model and flight test results will be adopted as criterion. If fewer inputs could result in approximation to baseline, main influencing factors should be limited within these fewer input

FIGURE 10 Comparison between model and flight test result of P_{od}.

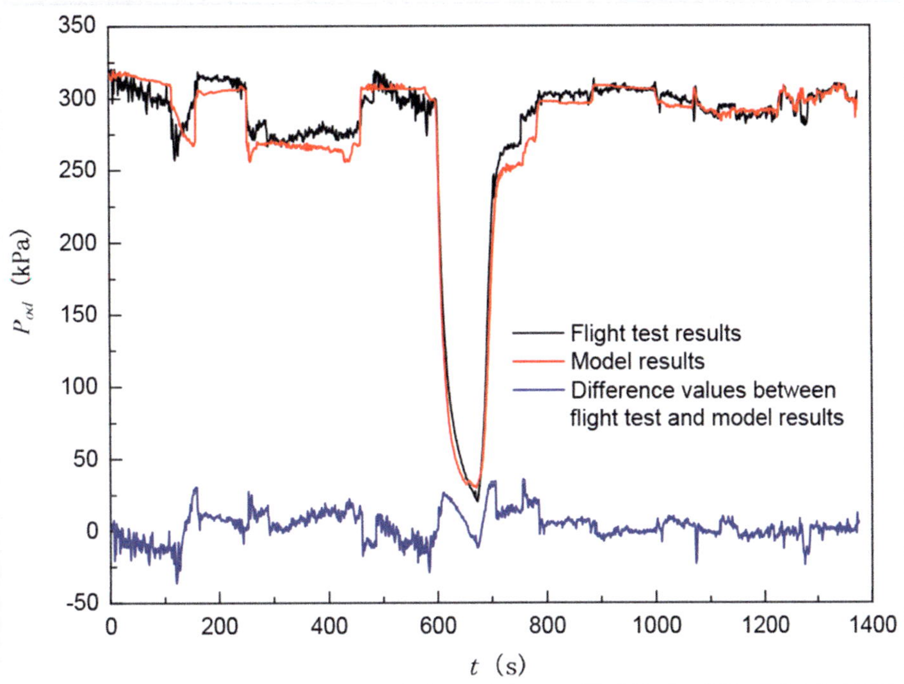

FIGURE 11 Comparison between model and flight test result of T_{or}.

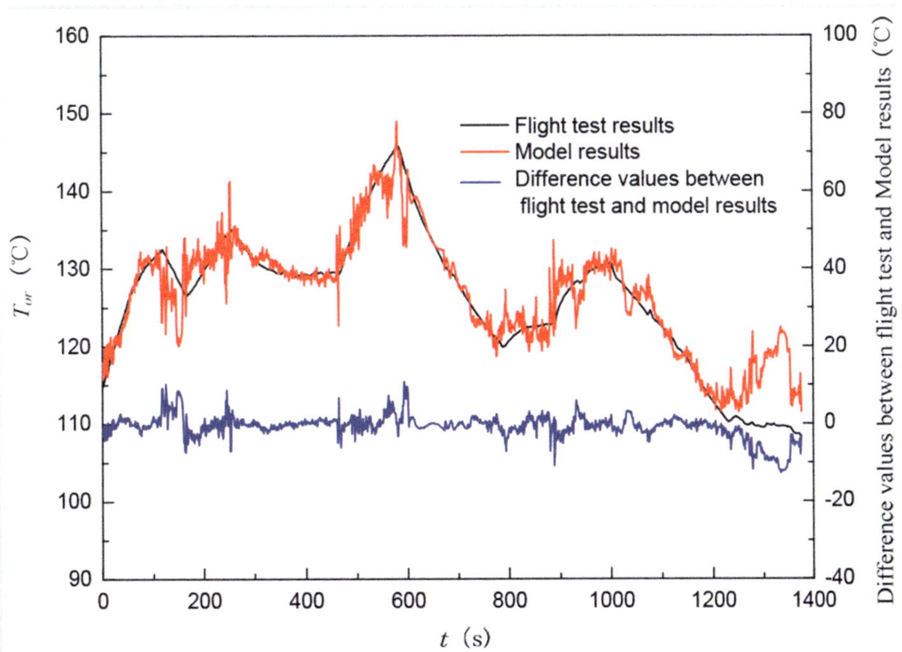

parameters. In this way, a method to determine the main influencing parameters of aeroengine lubricating system can be developed, which will help lubricating system flight test content design and malfunction prediction. The following context will discuss how main influencing factors of oil differential pressure P_{od} and scavenge oil temperature of rear bearing cavity T_{or} are determined.

By ANN, the number of input parameters has been reduced (for instance, if there are seven inputs, $C_8^7 = 8$ input groups can be produced; if there are six inputs, $C_8^6 = 28$ input groups can be produced, etc.), until the maximum difference value between model and flight test result is unacceptable. There are a total of $\left(C_8^7 + C_8^6 + C_8^5 + C_8^4 + C_8^3 + C_8^2 + C_8^1 = 162\right)$ different combinations of input parameters. Ultimately, the main influencing factor is the combination with least input parameters, and its maximum difference between model and flight test approximates baseline results.

Figure 12 is the distribution of difference value between model and flight test results of P_{od} with seven input parameters, which are *H*, *Ma*, *TAT*, *Ny*, *Nz*, *PLA*, and n_2, and the results are the closest to baseline results in all results obtained utilizing seven parameters as inputs. Figure 13 shows the distribution of difference value with five input parameters.

Figure 14 shows the distribution of difference value with three input parameters, including *Ma*, *Nz*, and n_2; and the difference range between model and flight test results is from −19.0 to 20.8 kPa. Utilizing the three parameters as inputs could obtain most approximate results to baseline results. Figure 15 gives a better comparison of Figures 12, 13, and 14. However, the results of the other three parameters as inputs are not as good as the parameter combination of *Ma*, *Nz*, and n_2. Figures 16 and 17 are the result of using the combination of *H*, *TAT*, and *Nx* and combination of Ma, *Ny*, and n_2 as input parameters, respectively; and the difference range between model and flight test results of the former is from −191.0 to 224.0 kPa, and the latter is from −51.4 to 27.6 kPa. Both results are unacceptable compared with the result for −19.0 to 20.8 kPa.

Two input parameters as well as one input parameter lead to unacceptable bias.

FIGURE 12 Distribution of difference values between model and flight test results of P_{od} with seven input parameters.

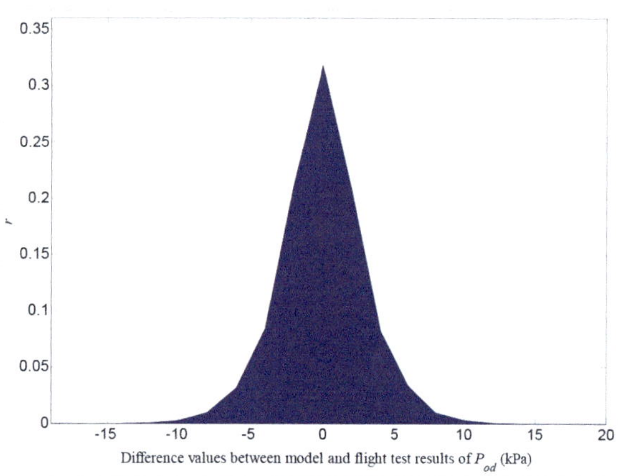

FIGURE 13 Distribution of difference value between model and flight test results of P_{od} with five input parameters.

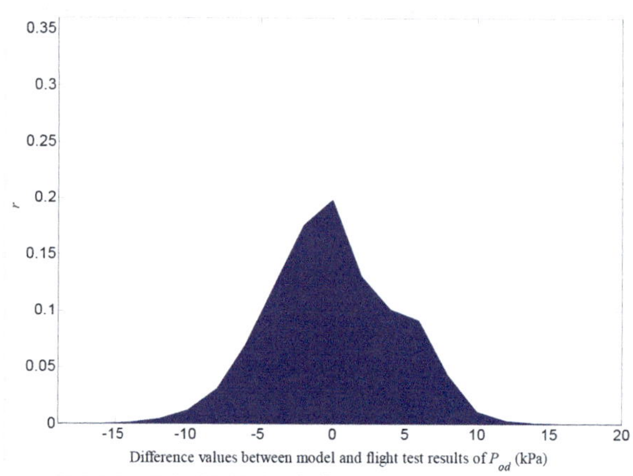

FIGURE 14 Distribution of difference value between model and flight test results of P_{od} with three input parameters (*Ma*, *Nz*, and n_2).

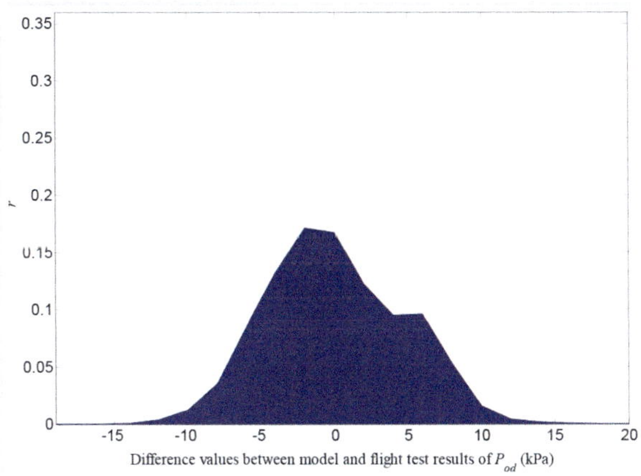

FIGURE 15 Approximation results of P_{od} with seven, five, and three input numbers.

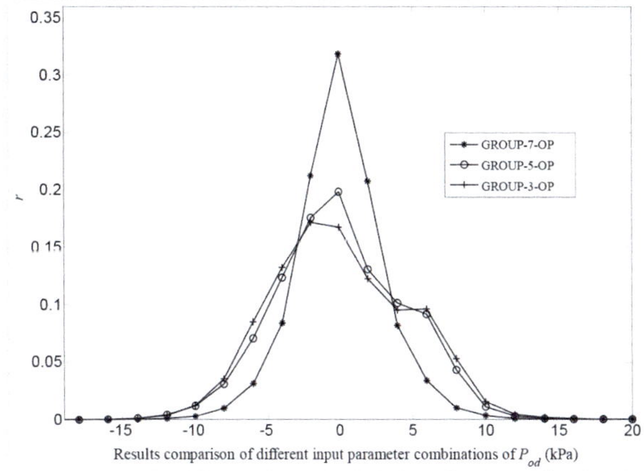

FIGURE 16 Distribution of difference value between model and flight test results of P_{od} with three input parameters (*H*, *TAT*, and *Nx*).

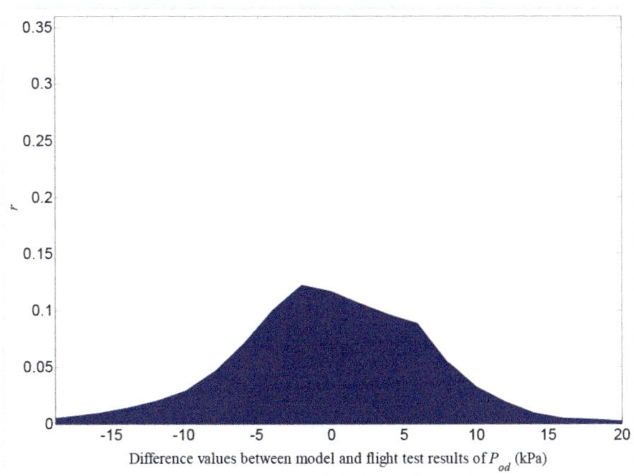

FIGURE 17 Distribution of difference value between model and flight test results of P_{od} with three input parameters (*Ma*, *Ny*, n_2).

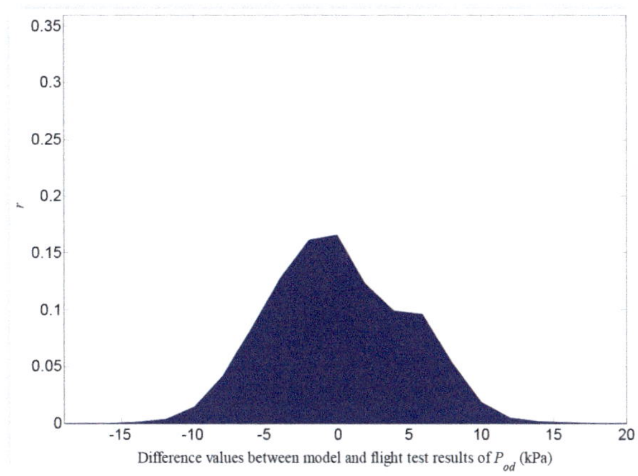

Figure 18 gives approximation between model and baseline of oil differential pressure with various input numbers, where GROUP8 stands for eight input (baseline input) parameters and GROUP_7_OP stands for seven input parameters, etc.

It can be summarized that the main influencing factors of oil differential pressure P_{od} are flight Mach number *Ma*, normal acceleration *Nz*, and high-pressure compressor speed n_2.

The main influencing parameters of T_{or} are also studied with the aforementioned methods. Distribution of difference value between model and flight test results of T_{or} with 7, 5, and 3 input parameters are respectively presented from Figures 19, 20, and 21. Figure 22 gives a better comparison of Figures 19, 20, and 21. Figure 23 shows approximation between model and baseline of T_{or} with various input numbers. It can be seen that the main influencing factors of T_{or} should be oil pressure difference P_{od}, *PLA*, and fuel-air pressure difference $P_{f\text{-}3}$, which represents fuel flow mass.

FIGURE 18 Approximation between model and baseline of P_{od} with various input numbers.

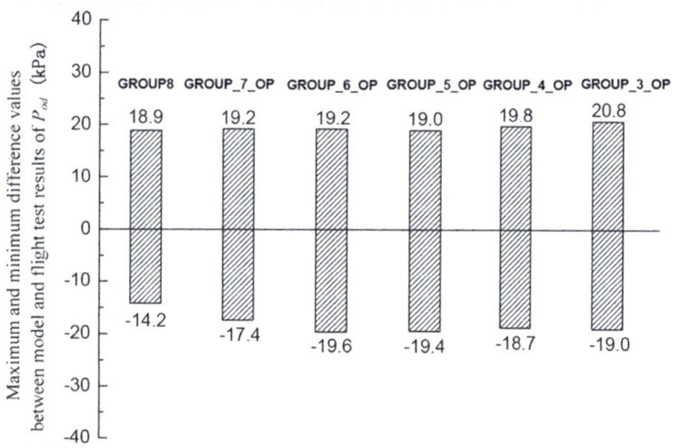

FIGURE 20 Distribution of difference value between model and flight test results of T_{or} with five input parameters.

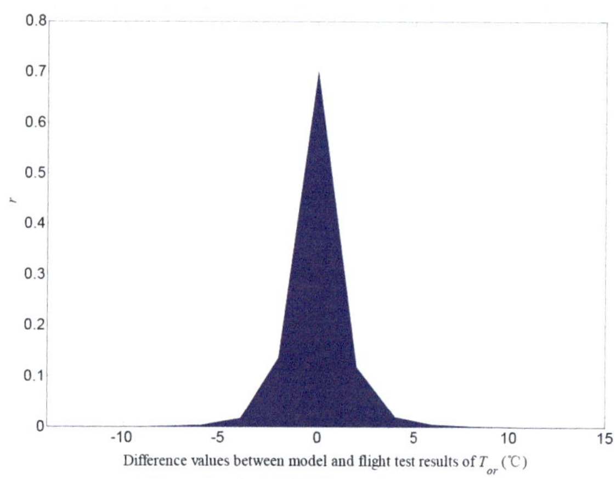

FIGURE 22 Approximation results of T_{or} with seven, five, and three input numbers.

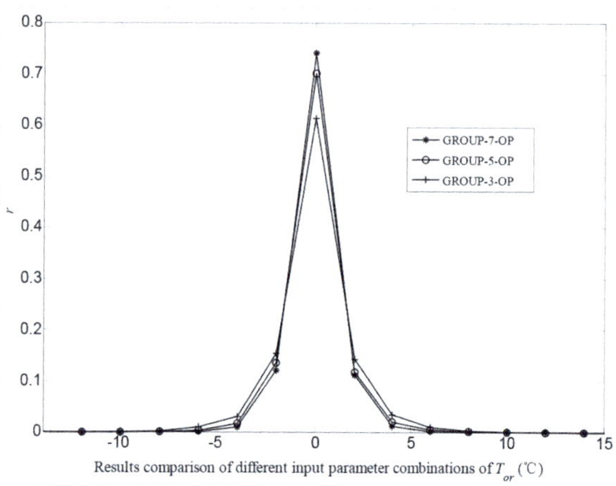

FIGURE 19 Distribution of difference value between model and flight test results of T_{or} with seven input parameters.

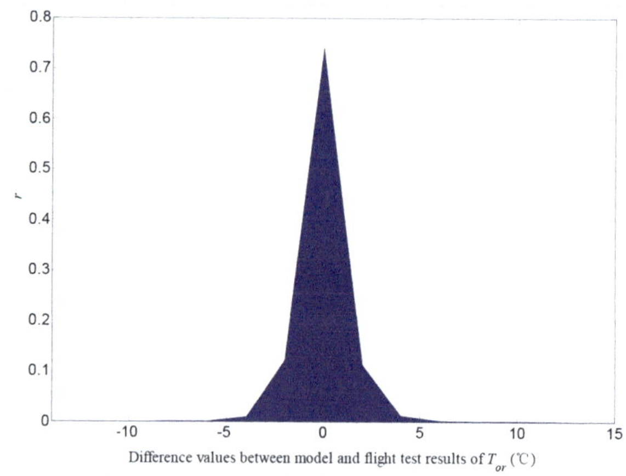

FIGURE 21 Distribution of difference value between model and flight test results of T_{or} with three input parameters.

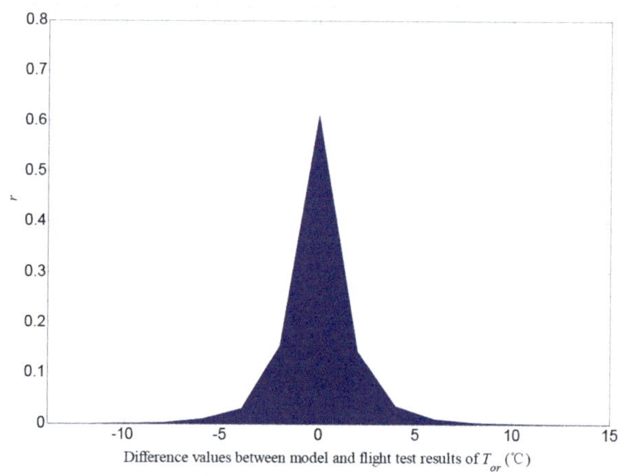

FIGURE 23 Approximation between model and baseline of T_{or} with various input numbers.

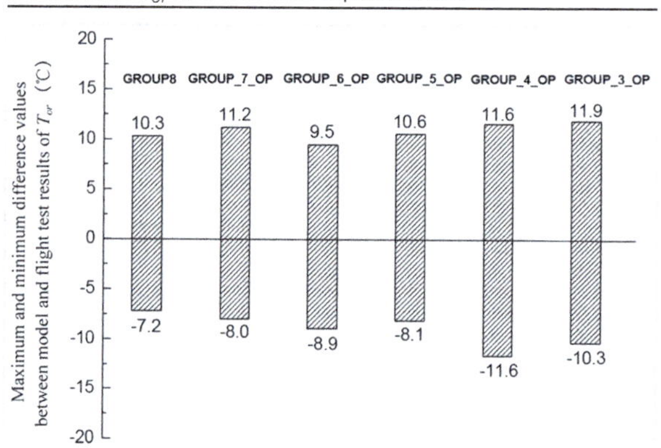

Summary

1. Basing on a large number of flight test data, models of six lubricating system parameters in all engine operation settings and whole flight envelope are established by utilizing ANN. Feasibility and effectiveness of the presented modeling method are demonstrated by the good agreement between model and flight test results.
2. Oil model is applied to an aeroengine lubricating system condition monitoring, and the coordination between calculating model and GDAS is accomplished. Comparison of model and flight test results in real-time monitoring comes true; thus online trend monitoring of oil parameters is implemented and applied.
3. By utilizing the lubricating system modeling method, an approach to determine the main influencing parameters of aeroengine lubricating system is developed. The main influencing factors of oil differential pressure are flight Mach number, normal acceleration, and high-pressure compressor speed, and main influencing factors on oil temperature of rear bearing cavity should be oil pressure difference, *PLA*, and fuel-air pressure difference.

References

1. Tony, G., *Gas Turbine Handbook: Principles and Practice* (Lilburn: The Fairmont Press, 2009) 94-100.
2. Royce Rolls, "*The Jet Engine*," (England: Technical Publications Department, Rolls Royce plc, 2005), 73-83.
3. USA Society of Automotive Engineers, Inc, "Guide to Oil System Monitoring in Aircraft Gas Turbine Engine," AIR1828B, Aerospace Information Report, 2005, 12-15.
4. Phillip, W., "Oil Monitoring Technology," *Proceedings of the IEEE Conference on Advancing Mission Accomplishment*, (Washington, DC: IEEE Computer Society, 1990), 523-528.
5. Narek, S., Jeremy, A., and Olivier, D.W., "Multistate Analysis and Optimization of a Geared Turbofan Engine Lubricating System," AIAA-2012-5637.
6. Ma, Z. and Cheng, L., "Steady Pressure Model of the Aeroengine Lubricating System," *Journal of Aerospace Power* 19, no. 3 (2004): 398-401 (in Chinese).
7. Zhou, Q., Liu, B., and Cheng, L., "Algorithmic Construction for Aero Engine Lubricating System Steady Model," *Aeronautical Computer Technique* 35, no. 2 (2005): 98-104 (in Chinese)
8. Yu, L. and Li, G., "Simulation of Pressure and Flux of Oil Supply System for an Aeroengine," *Aeroengine* 35, no. 6 (2009): 13-16 (in Chinese).
9. Su, L., Liu, Z., and Lu, Y., "Pressure Simulation of Lubricating System for an Aeroengine," *Science Technology and Engineering* 12, no. 1 (2012): 97-101.
10. Hashim, H., Haron, M.A., and Osman, F.N., "Artificial Neural Networks Based Algorithm for Identifying Engine Oil Parameters," *Second International Conference on Computational Intelligence, Communication Systems and Networks*, 2010, 244-249.
11. Zhang, B., Yan, J., and Yan, H., "Research on Oil Ferrography Analysis Fault Diagnosis of Diesel Engine Based on BP Neural Network," *Second International Symposium on Intelligent Information Technology Application*, 2008, 519-524.

12. Li, Y. and Zhong, S., "Condition Monitoring of Aeroengine Lubricating Oil System Based on Recurrent Process Neural Network," *Lubricating Engineering* 31, no. 9 (2006): 29-32 (in Chinese).
13. Gao, F., Zhu, F., and Wu, B., "Support Vector Machine-Based Prediction of Metal Content in Rotating Machinery's Lubricating Oil," *Aeronautical Computer Technique* 34, no. 4 (2004): 17-19 (in Chinese).
14. Fan, G. and Ma, D., "Probabilistic Prediction Method for Aeroengine Performance Parameters Based on Combined Optimum Relevance Vector Machine," *Acta Aeronautica et Astronautica Sinica* 34, no. 9 (2013): 2110-2121 (in Chinese).
15. Wei, X., Li, Y., and Wang, S., "Aeroengine Lubricating Monitoring Analysis via Support Vector Machines," *Journal of Aerospace Power* 19, no. 3 (2004): 392-397 (in Chinese).
16. Simon, D.L., "An Integrated Architecture for On-Board Aircraft Engine Performance Trend Monitoring and Gas Path Fault Diagnostics," NASA/TM-2010-216358, 2010.
17. Simon, D.L., "Propulsion Diagnostic Method Evaluation Strategy (ProDiMES) User's Guide," NASA/TM-2010-215840, 2010.
18. Zhu, J., "The Development Trend of Aeroengine Control," *Aircraft Design* 23, no. 1 (2003): 58-61 (in Chinese)
19. Chen, Z. and Li, Q., "Monitoring Parameters Forecasting and Faults Recognition," *Journal of Aerospace Power* 9, no. 4 (1994): 419-421 (in Chinese).
20. Zhang, R. and Xie, W., "Research on Aeroengine Lubricant System Fault Diagnosis Based on Wavelet Neural Network," *Aeronautical Manufacturing Technology* 6 (2009): 85-89 (in Chinese).
21. Mercer, C.R., Simon, D.L., and Hunter, G.W., "Fundamental Technology Development for Gas-Turbine Engine Health Management," NASA-TM-2007-0022364, 2007.
22. Stephen, O.T. and Singh, R., "Advanced Engine Diagnostics Using Artificial Neural Networks," *Applied Soft Computing* 3, no. 3 (2003): 259-271.

Contact Information

Author's name: Mingming Ma
Author's Institution: Institute of Aircraft Powerplant, Chinese Flight Test Establishment
Mailing address: Institute of Aircraft Powerplant, Chinese Flight Test Establishment, Yanliang District of Xi'an, Shaanxi Province, China.
Postcard: 710089
mmm_2007@live.cn
Telephone number: +86 181 9102 4939

Definitions/Abbreviations

H - flight altitude
Ma - flight Mach number
ϕ - roll angle
Nx - longitudinal acceleration

Ny - lateral acceleration
Nz - normal acceleration
n_2 - high-pressure compressor speed
P_f - combustion fuel pressure
P_3 - high-pressure compressor discharge pressure
$P_{f\text{-}3}$ - *Pf- P3*, fuel-air pressure difference between combustion fuel pressure and high-pressure compressor discharge pressure
P_{os} - oil supply pressure
P_{od} - oil differential pressure
T_{os} - oil supply temperature
T_{om} - scavenge oil temperature of middle bearing cavity
T_{or} - scavenge oil temperature of rear bearing cavity
T_{oi} - integrated scavenge oil temperature
S_t - total sample size of flight test data
S_d - distribution size of flight test data
r-Sd/St - ratio of distribution
ANN - artificial neural network
CFD - computational fluid dynamics
GDAS - ground data acquisition system
PLA - power lever angle
SVM - support vector machines
TAT - total air temperature

about the author

Dr. Ravi Rajamani is an independent consultant who has accumulated many years of experience in the area of aerospace propulsion and energy, specifically in data analytics and model-based methods for controls, diagnostics, and prognostics. He has three books to his name including *Electric Flight Technology: The Unfolding of a New Future*, published by SAE International Press, many book chapters, journal and conference papers, and patents. Prior to his current job, Ravi worked at Meggitt, United Technologies Corporation, and the General Electric Company. He has a BTech from IITD, an MS from IISc, a PhD from the University of Minnesota, and an MBA from the University of Connecticut. He is active within various SAE technical committees dealing with PHM and electric propulsion. He is also active in the PHM Society, serving on its board of directors. Ravi is Visiting Professor at Cranfield University. He is the editor-in-chief of the *SAE International Aerospace Journal*; and has been elected a fellow of SAE and of IMechE in the UK. In 2018 he received SAE's Forest R. McFarland Award.